FURTHER TALES OF ALYIE STARSTRIKER

Joseph Di Rienzi

Further Tales of Alyie Starstriker
By Joseph Di Rienzi

Published by TJT Press,
8567 Harris Avenue, Baltimore, MD 21234

"Further Tales of Alyie Starstriker" continues the saga of Alyie Starstriker, now High Governor of the Sidereal Federation in the 27th century. Alyie faces both personal and professional challenges as leader of 110 billion humans who now inhabit over 30 planets in a nearly 100 light-year diameter sphere of space. She faces threats such as an invasion from a predatory alien civilization, an ancient sacrificial cult, giant killer ants and governmental treachery at the highest levels. With Alyie's eternal partner Jaynis, their twin children – Tessa and Tomus, and colleagues, they meet each crisis with intelligence, strength and courage.

About the Author

Joseph Di Rienzi, is a physicist and Professor Emeritus at Notre Dame of Maryland University. He received his Ph.D. at Polytechnic Institute of N.Y.

For over twenty years he worked as a Research Scientist Contractor at NASA Goddard Space Flight Center. His research specialties are antimatter-matter interactions and the foundations of quantum mechanics.

He is the author of numerous physics publications, the multivolume "History of Thoroughbred Racing in the United States" and "Tales of Alyie Starstriker". He has two grown children, David and Sara, and lives in Maryland with his wife and counterpart, Joyce.

Dedication

To Joyce,
my partner, confidant and lover.

Table of Contents

The Reign of Alyie Starstriker - Part I

When the stars threw down their spears
And water'd heaven with their tears:
Did he smile his work to see?
Did he who made the Lamb make thee?

Tyger Tyger burning bright,
In the forests of the night:
What immortal hand or eye,
Dare frame thy fearful symmetry?

The Tyger by William Blake

Prologue

As the spacecraft Alux was hurled out of the hyperknot, all aboard hissed at the shock of returning to normal space with stars and galaxies surrounding them. At what served as the command deck, orderlies took their customary seats to resume their duties. Captain Thu?glr Kap?stk put its gloves back on over its hairy sweated-drenched digits and firmly grasped the pilot's control stick.

In the Rhilun language, First Navigation Officer Glib D?rhim announced the star system they were now traveling through. However, what gained everyone's attention was First Communication Officer Ben?ib Akal?een's declaration that a squad of spacecraft had been detected in the vicinity of the star we call Tau Piscis Austrini.

Captain Kap?stk in language punctuated with glottal stops growled, "Do you sense living organisms in these ships?"

"No Captain," Officer Akal?een replied, "They are all autonomous and appear on a reconnaissance mission of some sort."

The Captain narrowed its iridescent eyes, stroked its whiskers and purred softly. "Put a track on them and determine their home world. We need to assess their strength before we make this space part of the B?alamian Empire's hunting grounds."

Chapter 1.

High Governor Alyie Starstriker entered the Sidereal Federation Headquarters building on planet Teagarden with her ever present troop of Elite Guardsmen. All those inside stopped their activities, stood in rough attention and greeted her with nods and bows as she passed.

Alyie was outfitted in a long tan riding coat that was draped over her shoulders. Underneath, she wore a double-breasted, soft synthetic leather white jacket replete with medals and ribbons signifying her high office. A burgundy cravat sported the High Governor's insignia pin. Her snowy white breeches were tucked into tall burgundy "leather" riding boots with spurs. She had her laser pistol snug in her black holster belt and swung her riding crop lightly to and fro.

Acknowledging the respect her subordinates were giving her, Alyie tapped her cap with her riding crop. As she strove forward, a woman and a young girl stood in Alyie's path. The woman asked if her daughter could meet the High Governor. Alyie turned to the captain of the six guards attending her, who promptly ordered one of his men to do a body scan on the two strangers. Nodding to Alyie that they were free of any harmful devices,

Alyie knelt on one knee in front of the youngster, took off her right burgundy leather glove and allowed the child to kiss her signet ring. When the girl looked up with her bright green eyes, Alyie gently touched the child's cheek with her left gloved hand and smiled warmly.

The High Governor and her retinue then boarded the lift to her office/bedroom suite on the 15th floor. Alyie had just returned from an exhausting day that was highlighted by a visit from the Emir VII of Mongoti. The latter, who saw himself as a master horseman, was indulged as Alyie arranged for a tour of the military facilities on Teegarden on horseback.

Accompanying the Emir over the gallops, Alyie enjoyed the ride, although the Emir's insistent flirting became tiresome. Being polygamous he saw no contradiction in Alyie and him beginning a tryst despite Alyie's nearly 14 E-yr. eternal commitment to Dr. Jaynis Bottombrook. After eventually accepting Alyie's refusal as final, the Emir dined with her "Mongoti style" in a makeshift tent that Alyie's adjuncts had erected in the middle of what served as a desert on Teegarden.

Reflecting on the Emir's amorous attempts, Alyie groaned to herself that with a formal dinner "tomorrow evening" in the Emir's honor, she might have to go through the same "cat and mouse" dance she did today. (The ride and the Emir's affection did arouse some feelings in Alyie but for a different partner.)

After Alyie took off her gloves and cap, one of the Elite Guards slipped off from her shoulder the dusty trench coat, removed Alyie's spurs and gave her boots a quick cleaning. The rest of the troop made a thorough security inspection of the suite. Captain Ledlow Foworth, upon expressing his satisfaction that all was well, was thanked by Alyie for his service. The Elite Guards left the High Governor's suite, but, as always, two were stationed as sentries just outside. Alyie made a quick check of the time and realized that she had a lot of work to accomplish before the evening would reach a climax.

Chapter 2.

The nighttime shades had darkened the interiors of Teegarden so that it appeared to be the middle of night. (Due to the planet's synchronous rotation around its star, nighttime needed to be simulated indoors.) Alyie Starstriker unbuttoned her jacket and removed her cravat, setting them both on a side board. Stripped to the top half to her white tank top, she kept on her breeches, belt, laser pistol in its holster and newly shined boots.

She stood opposite her dressing mirror in her opulent bedroom suite that had been built adjacent to her office after she assumed the High Governorship of the Sidereal Federation almost four E-yr. ago. Peering into the mirror, Alyie saw a woman approaching 40 E-yr, in the prime of life. With age, her beauty had only deepened as life experiences provided a gravitas that made her only more attractive. Over her left breast was the scar from a wound that nearly claimed her life, but her skin was still creamy and supple.

Stroking her lustrous golden red hair that she had let down, she reflected on her children, Tessa and Tomus, now nearly nine E-yr. old, and Jaynis Bottombrook all residing on Earth. She hadn't seen them in person in four E-m. Alyie had promised to travel to Earth, but new situations kept detaining her in Teegarden.

They had a family vacation planned next E-m. which she silently swore she would not miss. An overwhelming sense of fatigue gripped her as the day had been a relentless series of meetings, hyperlink conferences and, of course, playing host to the Emir.

Today had also been her bi-weekly hyperlink session with citizens of a particular Federation system who be afforded the opportunity to address the High Governor with their concerns. This week's world was Nespar, where about 15 E-yr. ago Alyie helped defuse a possible civil war.

Moving over to her 18th Earth-century oaken desk in the bedroom suite, Alyie took a bottle of liqueur that the Emir had presented her as a gift. He claimed the grains were infused with certain flowery petals that would serve to relax the mind and sooth the heart. She placed two thin goblets on her desk and filled each with the elixir.

Using her implants, Alyie projected a holographic image of a proposed advanced starbomber that Space Command was seeking approval for. As she was rotating the design to examine it from all positions, she sensed another presence in the room approaching her.

Sensing familiarity from his scent and the clicking sound of the quantum lock on the secret entrance that allowed him to slip unnoticed into her sanctum, Alyie soon felt his warm lips against her neck, as she reached out her left arm to touch the hairs on his chest without even turning around. On reflex, she turned on the security field so that all outside communications were shut off.

"Working late as usual, I see."

"This design is much more complex than earlier models, and it is très expensive. We need to conduct an extensive analysis of the cost/benefit ratio of these bombers."

"Yes, Madam High Governor, but that can be accomplished tomorrow."

Alyie turned to face Colonel Reg Jones, Commander of the Panthera and the Director of Sidereal Federation Security. He was dressed in civilian clothes wearing an unbuttoned orange long-sleeved shirt, black loose-fitting pants, and sandals. His curly brown locks were now speckled with gray, as was his chest hair. Reg's laser pistol could be seen protruding from his shoulder holster.

"I find waiting for you is more endurable if I delve into work," said Alyie before planting a wet kiss on Reg's lips.

"I had a working meeting in my office with my staff, so I could not use the portal until I was sure all my subordinates had concluded their duties. To think, here I am with you, and outside there are guards and a security system protecting you against intruders."

"We have certainly found a novel use of quantum security," Alyie mused.

"My great fear is that someone can use it to get directly to you."

"Since you and I are the only ones who know the passcode, I think our secret 'tunnel' is safe," Alyie concluded as she offered Reg a goblet.

"Here's looking at you, my gracious Red Queen."

"Bottoms up, my valiant Black Knight," said Alyie, toasting Reg as she emptied her glass filled with the elixir.

"Speaking of 'bottoms up,' it is time for bed," Reg answered as he swallowed the liqueur, "but let's finish your disrobing."

Alyie promptly pulled her tank top over her head. Standing directly in front of Reg in her boots and breeches, she grasped his shoulders firmly. As Reg knelt to gently pull her boots off, Alyie planted first her left boot in his crotch, making a mischievous grin. After repeating the same gesture with her right boot (and getting the same response), Alyie unbelted her laser hostler and quickly pulled her breeches down. Reg lifted her up in a burst of passion. Alyie, who was a tall, athletically built woman, nevertheless felt light as a Martian snowflake in Reg's chocolate-colored muscular arms. He placed her carefully in the soft bed and soon joined her as he quickly disrobed and disarmed. As they were entwined, all thoughts of the twins, Jaynis, and the 110 billion citizens of the Sidereal Federation that Alyie ruled dissolved. As she and Reg reached a crescendo together, Alyie cooed in rapturous ecstasy.

Chapter 3.

As the shades began letting in the sunlight to give the appearance of the start of a new day, Reg stirred in the bed. Looking at Alyie asleep by his side with her gorgeous hair splayed upon her pillow, her arms spiraling in opposite directions, and her beautiful pink breasts and comely freckled face, he thought she was indeed one of nature's greatest creations. He was content in the place he stood in her life. He wanted to serve her in any way he could, whether that was as protector, confidant, or lover. Reg understood that her life with Jaynis and the twins was important, and he wouldn't do anything to jeopardize her happiness with them. Realizing he had to return to his own domicile, he got out of bed and quickly dressed. Before leaving, Reg stood over Alyie, kissing her lightly on the cheek, saying, "Sweet dreams, Madonna."

Chapter 4.

When Alyie arose later that "day", she was somewhat disappointed to find Reg gone, but she knew it was for the best. Showering to remove all traces of yesterday's events, she had her breakfast brought into her bedroom by her autobot. As she ate, she heard human activity in the anteroom to her suite. Alyie's adjunct, Ms. Hilda Burton, had come in and was conversing with one of the Elite Guards posted outside the office. Hilda was a longtime aid to Felicity Erickson, the previous High Governor. After Felicity's tragic death, Alyie retained Hilda in her service. Alyie and Hilda were not the most compatible workmates, given the difference in age and Hilda's expectations based on her service to Felicity. Still, Hilda had vast experience, and she was loyal to a fault to the person she worked for. She could be trusted to keep all she read, saw, and heard confidential.

"Good morning, Hilda. How does this day find you?" Alyie ventured to ask as she peered out of her office door to the anteroom in her dressing robe.

"I am fine, Alyie. How did your jaunt with the Emir go yesterday? Did he remember to keep his hands on the reins?"

"Part of the time, Hilda. Thank goodness I don't have to fend him off most of today."

"Yes, Alyie, he and his aides have several meetings with officers of the Common Exchange and the Sidereal Federation Bank."

"I am glad I had the directors of these institutions come to Teegarden instead of having the Emir and his staff travel to Earth. The fear is that he would insist I travel with him."

"Speaking of Earth, I want to remind you that you have a scheduled video hyperlink session with Tessa and Tomus soon."

“Oh, thanks, Hilda. I nearly forgot that. I better get dressed. I don’t want the twins to think I have become slovenly.”

“Hardly, Madam High Governor. I need to also bring to your attention that you have a High Governor’s Board Meeting in one E-hr.”

Yes, we will be examining the first data returned from the exploratory mission sent to the Tau Piscis Austrini System. I am eager to see what’s been found.”

Also, Alyie, you have an appointment with the autobot hairdresser late this afternoon and the autobot dress designer to outfit you for the formal dinner tonight.”

“Ugh. Outfit indeed,” said Alyie. “It will be more like arming me for defensive combat.”

“Also, the pigmentologist wanted to know if you wanted to change your skin color for tonight’s event.”

“No, I will pass on that.”

Leaving Hilda to her duties, Alyie finished dressing. For the meeting, she wore a tailored gray and thin red striped jacket and skirt suit. Underneath the jacket was a sleeveless satin, ivory-colored shell blouse adorned with a pearl necklace. On her feet, Alyie put on tall black high heels that she had gradually become comfortable wearing.

Initiating the hyperlink session with her implant, Alyie saw the smiling faces and light-colored hair of Tessa and Tomus pop up.

“Hello, Mommy, they chimed in unison. When are you coming to be with us?”

“Soon. We are all going to go on a vacation to visit Queen Revita on Kanva next E-m. She has promised to take us solar sailing, as well as horseback riding and site seeing. We will also celebrate your birthdays. It will be great.”

"We want you with us now," pleaded Tomus blinking his bright blue eyes.

"I can't get away right now, but a month will go by quickly. Tell me what you are learning with Taki-3."

After extracting, with some difficulty, the lessons they were engaged in math, reading, language, writing, and history, Alyie asked, "Can I speak to Daddy?"

"He is at the Université," replied Tessa, whose hair was becoming more reddish.

"Oh yes, I forgot. I will contact him there. I love you both, and I will speak to you next E-d. Goodbye."

A wave of guilt swept over Alyie. She was badly falling down as a mother and wife. It was not just her liaison with Reg. That was just a symptom, not the cause. Her role as High Governor was all-consuming and incredibly intense. Being with Reg provided her with some release. She wondered if Jaynis knew or at least suspected. If he did, he was being incredibly understanding, which made Alyie feel even more wretched for her actions. Jaynis and she had discussed moving the whole family to Teegarden, but that would disadvantage Jaynis in terms of his work, and realistically, how much time could she give her family? Alyie was pinning all on this trip to Kanva to right the rocky Starstriker-Bottombrook family ship. Perhaps, she was asking too much, but that was the only hope she could see on the horizon.

Chapter 5.

Leaving her office suite, Alyie found Reg waiting outside to escort her to the Board Conference Room. He was dressed in the uniform Alyie designed - a double-breasted black jacket, trim blue pants with a black stripe on each leg tucked into highly polished boots. With his laser belt holster around his trim waist, he looked every centimeter the soldier and gentleman Alyie knew him to be.

"Good morning, Madam High Governor. I hoped you slept well last night."

"It took me a while to unwind, but once I found my rhythm, it was delightful. How about your night, Colonel?"

"The same, Madam High Governor."

"I see from the auto-scheduler on my implant you want me to meet with an engineer from the Safety and Security R&D Group after the Board Meeting. Is there a sense of urgency, Colonel?"

"She has been developing, under my orders, a new defensive wearable gear for you. I would like you to take a look at it."

"Ok, but you know how style conscious I am."

"No worry, it is transparent."

At the High Governor's Board Meeting, in attendance were Alyie, Reg, Admiral Steely Cranshaw, Head of Space Command and Alyie's former mentor, General Ivan Presoff, Ground Forces Commander, and the 10 elected members from all sectors of the Sidereal Federation. Also invited was Dr. Sonja Benita, Director of the Federation's Scientific Research Institute.

The first order of business was the Quarterly Financial Report which enumerated revenues generated by the 30 odd worlds populated in Human Space through the Common Exchange and the enterprises of the Sidereal Federation Bank. Alyie noted that "although the report identifies continued growth in revenue, it projects an eventual flattening due to market saturation. Rather than increase the rates we charge member planets to do business in the Federation, I believe we need to expand the Sidereal Federation through further human settlement of space, investing long-term in building new habitats for people to live and prosper. This is the reason I approved the Mission to Tau Piscis, Austrini. Dr. Benita has kindly consented to join us today for an update. Dr. Benita, please."

"Thank you, Madam High Governor and Board Members. The drone constellation to the star system Tau Piscis Austrini is still fully operational and is sending data it has recorded through the hyperlink knots we had located en route. To summarize, Tau Piscis, Austrini is a yellow-white main sequence (F6 V) star about 60 l-yr. Away from the Sol System. It is slightly bigger and more massive than Sol but cooler. Younger in formation to the Sol System, the Tau Piscis, Austrini system has formed a debris disk surrounding the primary star that the probes have been able to identify three solid planet-like objects that may be suitable for habitation. Further inspection will use the algorithms aboard the drone constellation to provide a 'best candidate' biosphere to initiate terraforming."

"Excellent, Dr. Benita. Do the Board Members have any questions?"

"This system is much further than any other world we have populated previously," stated Mr. Rowan Manchester, an elected member. "Even if a body out there is selected for colonization, is it practical for humans to settle at the outer fringe of the Federation?"

"I would claim, Mr. Rowan," Alyie responded, "that the word 'fringe' is an elastic term. At one time, Teegarden, which is now the co-seat of the Sidereal Federation, was considered an outpost. The Federation needs to keep expanding if it wants to continue to prosper. Today's fringe is tomorrow's home world. If there are no other concerns, let us thank Dr. Benati and her colleagues at the Scientific Institute for their splendid work, and I am sure I speak for the rest of us in eagerly waiting for the result of their analysis. Once your program selects an object in that system, I want to move quickly to begin the development and migration process."

Chapter 6.

At the conclusion of the Board Meeting, Alyie consented to Reg's request to take her to the Safety and Security complex to look at a new device. Traveling in Reg's two-seater air-scooter, they made the quick trip to the building. Once inside, Reg directed Alyie to the research lab, where an engineer, Ja-Li Wang, was waiting for their arrival. She was dressed in a white lab coat, but from the bright blue color of her hands and face, it was clear she used the services of a pigmentologist. Bowing before the High Governor, Ms. Wang brought out what looked like a body suit that was made out of a fine mesh material. Attaching the suit to a manikin, the engineer activated it with a signal such that the suit repelled anything that was thrust at it, whether it was a knife, projectile, thermal bomb, or laser beam.

"The activation creates a repulsive electromagnetic (E-M) field. The suit can withstand any ordinary assault weapon other than a nuclear missile or an extremely high-intensity photon bomb."

"Is this like the force field that surrounds a spacecraft to deflect enemy assaults?" Alyie wondered.

"Yes and no, Madam High Governor," said Ja-Li, "They both use E-M waves to provide protection, but the mesh's field is highly local such that it is made for an individual, such as your Eminence. Also, it is designed to be a one-way directional shield. Objects will be repelled, but you are free to touch and hold anything without interference."

"Remind me, Colonel Jones, why do you think I need to wear this?"

"It is my job to protect you. We have the Elite Guards at your service, but I fear that when you are meeting the public, you are vulnerable to an assassination attempt as you greet people or meet with those who want secretly to usurp your rule."

“I thought I was popular and well-liked as High Governor. Do you have polls that indicate otherwise?”

“No, Madam High Governor. Your popularity makes you even more of a target. Human history is littered with charismatic figures who have been assassinated by those who are either jealous of the leader’s fame or have some sociopathic desire to bond by killing him or her. Trust me; you need to wear this.”

“Ok, I put this on my body. Can I control when I want it turned on and off? I can think of a few occasions when I would want it to be non-functional,” asked Alyie as she cast a sly smile at Reg.

“No worries, Madam High Governor,” answered Ja-Li. “Once you are fitted for the mesh, you can switch it on and off using a secure coded signal from your implant.”

“This might come in handy when the Emir tries to get too familiar while we are dancing tonight at the formal dinner.”

“I have a more direct way of stopping his advances, Madam High Governor,” was Reg’s agitated reply.

“Yes, I am sure that will look very manly and heroic, but I don’t want to have to arrest the Director of Federation Security for creating an interplanetary incident. No, for tonight, I will use my feminine guile to evade the Emir.”

Chapter 7.

Later that "day", the autobot coiffeuse, Coupe-1, came to give Alyie's hair a washing, remove any gray slivers, and put gold highlights in her red locks that would be worn fully down for dinner. When she was Regional Governor on Earth, Alyie had Nicole, a human, cut and style her hair. Under the much more utilitarian conditions of Teegarden, there were no human coiffeurs of any worth. While she waited in her dressing robe for her hair to dry and before the autobot couturière, Haute Couture-2, arrived, Alyie contacted Jaynis. He was meeting with a colleague, but he interrupted his conversation to take her call.

"Hello, Jaynis. How are you?" Alyie tried to present as cheerful an image as she could muster.

"I'm well," was Jaynis' polite but not overly enthusiastic reply. "I see you have had your hair done. What's going on at Federation HQ?"

"I am hosting the Emir of Mongoti. There is a formal dinner tonight for him and his delegation."

"Mongoti? That's in the Vega system. Isn't that the kingdom that had to be not so gently persuaded by none other than Seth Bullwark to join the Federation?"

"Yes, Jaynis, that was the current Emir's grandfather that Seth had to cajole with his Peregrine Falcon Group. Mongoti has always been a little tenuous in terms of its allegiance to the Federation. So this visit is to keep the Emir solidly in the fold."

"Why do you kowtow to this known despot?"

"We need Mongoti's resources. They have significant deposits of rare earth elements that are used throughout the Federation. A happy Emir makes one less supply problem for me to deal with."

"Whatever."

"How are you doing, Jaynis? Any recent research you would like to share?"

"I am traveling tomorrow to Mars to meet with my old colleagues from Chandrasekhar University. We are hosting a conference on gravitational quakes. Apparently, they can be quite sudden and dramatic. I will be there for about an E-w., not including the round trip time."

"Will the twins be going with you? When I spoke to them earlier, they didn't mention Mars."

"On the way to the Paris spaceport, I will drop Tessa and Tomus off at Alison's. She will take care of them while I am gone. She promises to give them a cultural education by bringing them with her to the Ministry of Arts and Culture. I think she has some trips planned around the Earth to further their sense of Earth's past. I complain a lot about the Panthera guards living here, but they are certainly useful in providing access and transportation to anywhere we want to go."

"That is very nice of Alison to do this. I am sure she is very busy with her duties as head of the Ministry."

"Alison has been great. I don't know what I would do without her. I have to go now. I have kept my colleague waiting when you called. Have fun at your dinner."

"Remember, Jaynis; we are all going to Kanva in about one E-m. It will be great to be together again."

"Goodbye, Alyie."

Feeling somewhat uneasy from her conversation with Jaynis, Alyie resolved to make things right and curtail her infidelity. Focusing on the present, she signaled for Haute Couture-2, who arrived presently for the fitting. After some time and much debate, Alyie and the autobot settled on a full length emerald-colored gown cut closely to the contours of Alyie's body. The gown had a scoop neckline and short sleeves. Alyie would wear the silver necklace her parents, Tanita Starbright and Alex Striker, had given her almost 15 E-yr. ago, which she matched with silver earrings. She accessorized the dress with long white gloves up to her elbows and cream-colored high heels. Looking in the full-length 360° mirror, Alyie saw a woman that for all the success she had achieved and all the happiness she had experienced, appeared, at the moment, unsettled.

Chapter 8.

After the night's events were over, Alyie was escorted back to her suite by her guards. After waiting patiently until they deemed her suite secure, Alyie bid them good night. She closed the outside door and walked haltingly across the office to her bedroom. The evening went as well as to be expected. She toasted the Emir and his delegation; he, in turn, saluted her as "The greatest and most beautiful High Governor in the history of the Federation". The trip was successful for the Emir in that he was able to secure a generous loan from the Sidereal Bank (which Alyie approved) and was able to set up more expedited trading protocols with the Common Exchange. With his mission accomplished, the Emir was on his best behavior. He was a perfect gentleman while holding Alyie as they danced around the ballroom. (The Emir did at the conclusion of the dance, whisper in Alyie's ear that he would be happy to continue their fruitful dialogue in his room if she so desired.)

As she began to remove her right glove with some difficulty, Alyie realized she had a little too much to drink. Feeling lonely at the moment, she had the impulse to call Jaynis but realized he would be sleeping. She reached for the bottle of liqueur the Emir had presented her and, filling a goblet, took a drink. Still unsatisfied, Alyie did what she had silently promised she wouldn't. She used her implant to send an encrypted quantum-entangled message,

"I need some help in getting my gown off. Can you come?"

Chapter 9.

As the interior of Alux was made transparent at all E-M wavelengths, providing a "cloak of invisibility" by reducing reflected waves from outside sources to a bare minimum, Captain Kap?stk growled to rouse a sleeping Ben?ib Akal?een, the ship's First Communication Officer,

"Wake up! I need your latest analysis of the origin of these drones."

Startled, the Communication Officer straightened up, declaring, "Sorry, Sir, the algorithm has just completed its processing. It has identified where these spacecraft are sending their signals. It is in the VCCI, system (author's note, i.e., Teegarden's star). From the amount of data and bandwidth of this stream, it appears the sentients have an advanced technological civilization."

"Can you tie into their communications and identify how far their reach spans and the strength of their military?" Captain Kap?stk inquired while cleaning its face with its tongue.

"The communications emitting from their home world extends to a sphere of radius of at least 30 l-yr., answered Akal?een. "Since the messages use hyperlink micro-knots, we can assume their spacecraft can travel through wormholes. I can't assess their military capability, but on this home world, there seems to be a lot of space traffic which would indicate a large fleet. I think we should assume they have a robust attack and defense system."

"Then it would be to our advantage," mused the Captain, "to use surprise as a weapon against them. First Navigation Officer, pick a soft target in their system as prey. We will maintain invisibility until we begin our conquest and claim glory for B?alam.

Chapter 10.

The plan was for Alyie to fly to Earth and spend a couple of E-d. with her family at their home in Montpellier near the Mediterranean Sea prior to their trip to Kanva. Reg Jones had commissioned the starbomber *Argonaut* as their transportation with a battalion of 500 strong Panthera to serve as escort. Alyie objected to the necessity of such a military presence, but Reg insisted, saying, "The last time a High Governor traveled to Kanva, she met with disaster. I want you to be protected as much as possible."

Alyie had issued an executive order that ceded much of the duties of running the Sidereal Federation to Reg as Director of Federation Security while she would be away. Ms. Ronica Richlieu was the designated Deputy High Governor, but by Federation law, she would only assume leadership if Alyie were dead or incapacitated.

In preparation for the trip, Alyie had been residing in the High Governor's mansion just a short distance from HQ. She remembered when she first assumed the High Governorship, she and her whole family lived there during the crisis with *Grendel*, the antimatter cloud. Things were frantic, but they were happy. She needed to bring everyone back to that mindset. After the trip to Kanva, she would convince Jaynis to relocate to Teegarden. She was sure she could find a position for him at the Scientific Institute. The twins would miss their schoolmates from Earth, but there were some families working for the Federation with similarly aged children that she and Jaynis could arrange get-togethers with. But, she thought, this is getting ahead of things. "Let's first have a great vacation on Kanva, find ourselves as a family again, and then make future plans."

On the day of her flight, she finished packing and dressed in her traveling outfit: A double-breasted white soft leather jacket and light brown cravat with the High Governor's insignia pinned on it on top of a white tank top pullover. Her tan breeches were tucked into high brown boots. After putting on brown gloves and fastening her cap, she placed her laser pistol securely in its holster belt that was fastened around her waist. Greeted at the door of the mansion by Reg, he draped her trench coat around her shoulders to keep her dry from the light rain that was falling.

Traveling to the spaceport in Reg's ground car, escorted front and rear by Panthera in aircars, the canopy of Reg's vehicle was open so citizens on the way to the airport could catch a glimpse of the High Governor. Arriving at the spaceport, Alyie and Reg jumped out of the car and, with much fanfare from those assembled, approached the ship.

As Alyie was about to board, she turned to Reg, "Take care and let me know of any developments."

"Don't worry about anything. Be with your family 100% of the time. I will mind the store here. Give my best to Jaynis and the twins."

"Thank you."

Chapter 11.

The not quite three E-d. flight to Earth was uneventful. After landing at the Paris spaceport, Alyie and a troop of four Panthera boarded an airjet for the short flight to Montpellier. When Alyie entered the chateau, she found some significant changes in the décor. Most startling was she was greeted by Alison Bernewithy, Minister of Arts and Culture. Not having seen her friend and former adjunct for some time, Alyie was somewhat taken aback that Alison had changed her hair color and style of clothing. Now a redhead with gold streaks, Alison was dressed in a green herringbone tweed sport coat that she wore over a yellow waistcoat. Underneath the vest was a starched button-down white shirt on which a yellow tie was knotted. Her tan breeches were tucked tightly into high black boots. With freckles that Alyie didn't remember her having before, sparkling blue eyes, and rose-colored cheeks, Alison was the picture of freshness, youth, and beauty.

"It's so great to see you; we have just returned from a great ride on the beach. The twins are changing their clothes in the Mud Room. Jaynis will be down in a minute. He had to return here before we did because a colleague needed some data that Jaynis had stored in a secure file at home. I am sure they will all be thrilled to see you."

"Jaynis joins you and the twins riding now?"

"Oh yes. It took some gentle persuasion, but he gallops right along with us now. I really think he looks forward to our rides together. Of course, we are not alone as your Elite Guards stay close by in their aircars."

While the two women were conversing, Tessa and Tomus made their entrance into the living room, and Alyie gave them both a huge hug.

“Mommy,” Tessa blurted out. “Aunt Alison has promised to take us to Newer York City when we return from Kanva. Isn’t that great!”

They had grown since she last saw them in person, and this made Alyie even more despondent over the time lost. After a while, Jaynis made his appearance. He seemed hearty and a little more filled out than Alyie recalled. He had evidently changed from his riding attire into his usual casual clothes of a pullover sweater, old worn pants, and soft shoes. His sandy-colored straight hair was uncombed, and he appeared to be growing a narrow blond moustache under his nose.

“Hello, Jaynis; it’s so good to be back. Things seem a little different here.”

“Oh, yes. I have given Alison permission to make this place more comfortable for me. I think she has done a great job. Don’t you?”

“If it makes you happy, then I am satisfied.”

After some awkward moments, when no one knew what to say. Alison chirped, “I need to be going back to the Ministry.”

“Not before you join us for lunch,” Jaynis insisted. “Alyie, Alison has arranged a welcome home meal for you. It would be desultory for her not to partake in this feast. I am serving some of my best vintage. Don’t you agree?”

“Of course, Alison will stay. As High Governor, I order it.”

Chapter 12.

On the day before the trip to Kanva, Alyie paid a visit to the Regional Headquarters complex in Brussels. This is where she worked as Governor before winning a seat on the then-High Council. Her successor, Hector Palmero, was nearing his fourth year in the position, and Alyie wished to speak to him. Informing Hector this would be more of a social rather than an official visit, Alyie was nonetheless greeted by what seemed to be all of the Regional staff members on the roof of the Headquarters Building when her airjet landed. After taking time to greet all those assembled individually, Alyie dismissed her guards and was escorted by Hector into the lift as they descended to the floor where her old office was located.

Inside the Regional Governor's suite, Alyie and Hector settled down on chairs separated by a small marble-topped table. Finger food and claret from the Bordeaux region were served by Hector's autobot. Hector had become grayer in his hair and beard in the four years since Alyie had last seen him, but he was still a strikingly handsome older gentleman. As usual, he was dressed immaculately in a light gray suit with blue pinstripes, brilliant white starched shirt, paisley patterned silk tie, and polished black shoes.

"There was no need to have a reception for me; I wanted this visit to be informal."

"Cariño mío Alyie, do you really think I could prevent my people from wanting the opportunity to meet you? This was a grand event for them. To be greeted in person by the beautiful Senora High Governor. They will be telling stories about this brief encounter for the rest of their lives. You don't know the impact you have had on the Federation."

"Speaking of impact, Hector, I want to congratulate you on the successful incorporation of the Ministry of Arts and Culture and the Common Exchange on Earth. You must have needed all your considerable diplomatic and administrative skills to accomplish these huge projects. I am so happy you are the Sol System's Regional Governor."

"Congratulations, indeed. You know quite well it was you, Alyie, who made these transfers as well as engineering the relocation of the Sidereal Federation Central Bank on Earth possible. Without your efforts on the High Council, the Earth wouldn't be thriving the way it is now.

"The one promise I have not fulfilled is declaring Earth the capitol of the Federation. I have debated this, but since all of the military and security systems are on Teegarden, I have refrained. I really don't want the Earth to be the epicenter for the Federation's armed forces. Since they are on Teegarden, I believe the headquarters of the Federation should also be located there, although having the High Governor moving to Earth would solve my own special problem."

"I sensed something was troubling you, and this visit was more than just social. Let me turn the security shield on so we can talk in confidence."

As Hector did this, he rearranged his chair so that he was now seated by Alyie's side. She uncrossed her booted legs and leaned toward him, putting her head on his right shoulder.

"Oh, Hector, I feel I have become unmoored. My family and I are drifting apart. This position is so demanding. I had no idea it would consume me soon much."

"Yes, it is a position of great responsibility and the way you have reorganized it places even more on your shoulders. But there is still time to find your equilibrium."

"When I just now returned to our chateau at Montpellier, I felt estranged from my children and Jaynis. I think they have counted me out as a prime family person and have made adjustments in their lives because of my absence."

"I understand you are traveling with them to visit your friend, Queen Revita Jadeite, on Kanva. This would be a great opportunity to re-forge your bonds with them."

"Oh, Hector, I think some of those bonds may be unrepairable. I have been unfaithful."

"Ah, si. Do you know High Governors have a history of promiscuity? Ji-hoon-Park, who you were acquainted with, was famous for the bevy of women that would make nightly trips to his suite. Your position is so intense that there needs to be some way to relieve the tension. Perhaps if your family moved to Teegarden, there would be no reason to look elsewhere for diversion."

"It is not that simple. This person I am involved with is not disposable. He is someone I value, trust above all others, and, on some level, I feel I need."

"Now I see the complication. In our time, there are many different arrangements that are accepted. You need to find the balance point, but it has to be agreed on by all people affected – your husband, your children, and your paramour. If a balance is unworkable, then you need to choose to discard something in your life, be it your high office, your family, or your lover. Not an easy decision but a necessary one. You cannot govern the Federation effectively if you are racked by internal turmoil."

"Thank you, Hector, for laying out the choices so clearly. I so value your advice and friendship."

"Senora, you are like a daughter to me, as well as the brightest light in the Federation's firmament. Anything I can do for you, I will."

Chapter 13.

The two E-w. journey to Kanva was enjoyable for Alyie and her family. The twins had never traveled that far before nor made two hyperknot jumps on the same trip. Captain Lenora Honeywine had given permission (with Alyie's assent) for Tessa and Tomus to be on the bridge for part of the journey, which they both took great delight in. Jaynis spent part of the time working on his research on gravitational quakes, but he and Alyie found time to discuss their children's lives and recall some of the good times they had shared together. Alyie wanted this to be a real vacation, so she avoided any talk about her duties as High Governor. After about an E-w, both had become more comfortable in each other's company and resumed their intimacy which was excitingly unconventional during a hyperjump.

As the *Argonaut* entered the Trappist-1 system, the blue orb of Kanva came almost immediately into view. The closeness of Kanva to its cool red primary star (class M) generated a temperate climate on the sunlight side. The enormous Andaman Ocean was partly responsible for an active hydrologic cycle, and coupled with its heavily oxygenated atmosphere, Kanva featured lush tropical regions on the rather small land masses. (Of course, the planet's synchronous four E-d rotation produced perpetual daylight on one side and darkness on the far side.)

As the starbomber touched down at the spaceport, a formal reception was there to receive them. Dispensing with protocol, Queen Revita Jadeite was the first person to greet them when Alyie and her family stepped out of the spaceship.

"Welcome to Kanva. I am so happy you have come to spend your time off with me." Revita was a petite dark haired woman with almond-brown eyes and a beautifully sculptured face. As befitting a queen, Revita was dressed formally in a forest green jacket with gold braids and a brace of medals pinned over her breasts, identifying her as head of the Kanvaian military. Her family's ancestral long slender sword hung by her side with a gold-edged hilt. The Princess' white dove-colored breeches were cut off by black leather highly polished boots that came over her knee. On top of her raven-colored hair was a black felt beret fastened by an enormous jade pin, the sigil of her family.

The two schoolmates hugged tightly and warmly. "I am so happy to be with you," Alyie cried, nearly close to tears. "You don't know how much this visit means to me."

Breaking free from each other, Alyie presented Tomus and Tessa, who showed that their practice paid off, with near-perfect bows to the Queen.

"Oh, let's not get into titles. I am just Revi. But look at both of you. How much you have grown."

"Thank you, Your Highness; I mean Revi," said Tessa.

"You are welcome. What do you both want to do when you are Kanva?"

"Solar sailing," the twins exclaimed in unison.

"Great. We can go tomorrow. I also hear someone has a birthday coming up. Who might that be?"

Both Tessa and Tomus shouted in unison, "We do!"

Revita then turned to Jaynis. "Welcome back, Dr. Bottombrook. The last time you were on Kanva was to rescue your wife from a crazed clone of my family. I guarantee you a much more relaxed visit this time."

"Thanks for hosting us, Your Majesty. We have been looking forward to this trip for some time."

"Jaynis, I also have a lot planned for Tomus and Tessa to allow you and Alyie to go on adult dates."

"Yes, Revi, that is much appreciated," Alyie jumped in. "Jaynis and I have not been alone in public in almost four E-yr. It will be a treat to go out together."

"I recommend Kanva's night side. Don't worry about security. We will provide all that's needed, and you will travel incognito. Just leave the twins to me."

Chapter 14.

While a welcome dinner was celebrated at the palace for Alyie, Jaynis, and the twins that “evening,” the Panthera feasted in a large erected tent adjacent to where they would be housed in barracks constructed on the parade grounds. (In addition, Reg Jones insisted Alyie maintain a twelve-strong Elite guard troop with her and her family at the palace.) While Jaynis and the twins were given a tour by Revita’s Lady-in-waiting, Ms. Rebina Ectot, Alyie, and Her Highness retreated to the Queen’s bedroom. Dispensing with their formal attire, the two women were in their undergarments and stocking feet as they sat propped on Revita’s enormous bed, sipping herbal tea from green flagons emblazoned with House of Jade crests.

“Tell me everything about your life, Revi. Not the official stuff, but what’s going on personally.”

“Not much. I have a good Leader of Parliament who runs the government efficiently, listens to my advice, and accepts my judgment as final. From what I hear, I am well-liked by my people. Of course, it helps that all know of my relationship with the High Governor. The assistance you provided after Kanva’s near brush with the antimatter monster helped re-settle our world and quickly brought back prosperity. People are wild about you. I had to use all my queenly power to keep your visit here as private as possible. If it were up to Parliament, there would be planet-wide celebrations honoring you with parades, assemblies, and festivals.

“Thank you for your discretion. I really want this to be a time for my family to be together. But I see you have artfully deflected my question. Let me be blunt. Why do you not have a Consort?”

“Oh, Alyie, I do not want to do that. But be calm; I have not been without some pleasure. You have met Rebina Ectot. Isn’t she a darling?”

“She may be all you need right now, but she will not help you extend the Jadeite line.”

“No, I don’t need a male Consort to do that. Parliament passed an edict saying as long as they approve of the gentleman I nominate, he can submit his sperm to be frozen. So at the appropriate time, my eggs can be fertilized to give me heirs. Right now, I am looking for suitable candidates.”

“This seems very clinical and prescriptive, but the Federation abides by The Act of Self-Determination, so who am I to oppose the will of the people of Kanva.”

“I know this may sound crass and hurtful to you, but I think Jaynis would make a great donor. Of course, I am sure he would consider this an act of betrayal to you.”

“Feel free to ask him. I am not sure he would consider this a betrayal since he has already been betrayed,” Alyie blurted out.

“What are you saying?” Revi gasped.

Alyie preceded to tell her friend about her affair with Reg Jones and the estrangement Jaynis had exhibited to her. She did hold back his growing interest in and reliance on Alison Bernewithy. That was too terrible for her to verbalize.

“That’s why this trip is so important. I need to rekindle the spark that has kept Jaynis and me together all these E-yr. On the way over, I thought I was beginning to make some progress.”

“And that is what you continue to do here. Leave it to me; this is going to be a second Sugarluna for you both.”

Chapter 15.

The sailing experience was great fun. The Starbrook family were outfitted in sailing clothes by Queen Revita. The twins looked very sea-worthy in their respective sailor girl and boy white and blue outfits. Jaynis had on a gray pullover shirt, loose dark blue fitting pants, and white boat shoes. Alyie wore a white short-sleeved shirt with red horizontal stripes, white ¾ length bright white pants, and red waterproof slip-on shoes with treads. She had a bright red kerchief around her neck and a large floppy white hat secured by a red ribbon under her chin. Revita matched her friend wearing the same outfit with green instead of red in her kerchief, stripes, and shoes.

The Queen, who had overcome her aversion to sailing as a result of her family's tragedy, was now a skilled mariner, as she used the ship's solar power to first get airborne and then, when sufficiently far from shore, had it descend onto the gleaming Andaman Ocean. Using the solar sails to power the electric engine, the *Virtuosity* glided smoothly through the waters. Tessa and Tomus, who had never been on a boat before, were thrilled with the experience as they stood at the bow to catch the breeze and leaned over the railings to catch glimpses of the exotic sea life. Revita also gave them both turns at the Captain's wheel, switching the autonomous steering to human control. Alyie and Jaynis reclined in the bright sunshine, enjoying the sky above and the sea below. When far enough out, Revita had members of the crew serve a festive lunch as they all sat on the foredeck.

Using the ship's imaging system to track schools of fish, Revita gave a quick training lesson in fishing techniques, and everyone tried their luck with the reels that were provided. In the end, Alyie was the leader hauling in 10 fish. Tessa bested her brother in catching five to his four. Jaynis could only haul in one, but it was by far the largest of the day. On the return voyage, Revita had *Virtuosity* alternately rise up and down, in and out of the water, giving a roller coaster ride that produced delighted screams from the twins.

Chapter 16.

The next Kanvaian "day," Alyie and Jaynis prepared for their night out on the small continent on the planet's dark side. Using pigments, their faces were colored in red and blue colors, and their garments were harlequin costumes of pullover shirts, tights, and soft fabric boots. These "disguises" would allow them to go unnoticed in the crowd and fit into the traditional Carnival Festival environment that was taking place on Kanva over the Earth fortnight. Revita convinced Alyie to dispense with her Elite Guards, saying they would look too obvious and attract unwanted attention. She commissioned four of the Royal Guards to dress appropriately and shadow Alyie and Jaynis throughout their sojourn on the Night Side. As for Tessa and Tomus, Revita would provide amusement for them rummaging through the hidden rooms and secret passageways in the palace.

Dropped off by an airjet on the outskirts of the Night Side capital of Nox, Alyie, Jaynis, and their entourage made their way in an airbus to the epicenter of the Carnival. Strolling through the narrow crowded streets, they were surrounded by dancers, jesters, and street performers. Alyie and Jaynis sampled the food, sang, and swayed to the music, both buoyed up by the general gaiety all around them.

Later in the evening, they found rest for their sore feet and balm for their parched throats at a small café away from the noise and excitement of Carnival. Alyie believed this was the time to talk, although she dreaded how the conversation would go.

"This has been great, Jaynis. It's good to be able to spend time together by ourselves."

"By ourselves, are you discounting all the revelers in the streets and shops? And let's not leave out the Queen's guards that are lurking somewhere in the shadows."

"Ok, so we are not absolutely alone, but we have had time to be just a couple. I love our children with my life, but their presence tends to demand our undivided attention."

"Are we still a couple, Alyie, or have others joined our little circle?"

"Oh, Jaynis, I don't know what to say. I know I have not been truthful to you, and I am deeply sorry for hurting you. I do love you as much as ever, and I want our family to stay together. Please believe me."

"How can we be a family with you on Teegarden doing what you are doing and me and the twins on Earth trying to make a life without a mother and a steadfast wife?"

"I don't want to make excuses, but being High Governor is demanding beyond anything I imagined. It has made me into a different person who does things and has needs I had never envisioned myself having. However, being back with you and the twins, I feel I am Alyie again. I know I am living a double life, but I want to find a way to reconcile these halves. I am asking a lot from you, but can you be patient with me?"

"Patience is hard, Alyie, but I am willing to wait a little longer. Our family is too important to destroy."

"Thank you, Jaynis." As Alyie leaned over to kiss him, a Code V warning was buzzing in her implant.

"It is from Reg Jones. I am sorry I have to answer this."

"Jones?" glared Jaynis. "Here he is again coming between us."

Alyie responded quickly to the distress signal. "Colonel Jones, what is happening?"

"Madam High Governor, we have just received a partial message that the G. Soffen Exobiological Institute on planet Hari in the star system you are in has been attacked by an unknown spacecraft and is putting out a distress signal. All indications are that the attackers are aliens with technology different than ours. Please protect yourself."

Chapter 17.

As Alyie and Jaynis were quickly making their way back to the palace, Alyie was receiving information about the assault on Hari and the identity of the attackers. Hari is a small rocky planet at the far end of the Trappist-1 system. It was set up three E-yr. ago as a research facility to conduct studies on possible life forms on newly discovered planets. The G. Soffen Exobiological Institute housed about 50 staff members. With housing and infrastructure, the population had grown to about 200. (The planetary surface being cold due to its distance from the cool red star and without an oxygenated atmosphere necessitated all living facilities be in indoor habitats in simulated Earth gravity.) Being at the far end of Human Space, Hari was provided with a robust defense system that was run autonomously.

The delayed hyperlink dispatches Alyie was receiving on her implant reported that the alien vessel was able to avoid detection until it entered Harian space due to its transparent hull. Once identified as an intruder, the defense system responded, but the alien spacecraft's thermal weapons were slowly eroding the heat shield protecting the Institute. From the latest message from the Institute's Communication Chief, it was not more than an E-d. before the shield would be down.

Patched into Reg Jones, Admiral Cranshaw, and General Presoff, Alyie pressed for an assessment of the enemy's strength and intent.

"Do we know anything about the spacecraft's design, propulsion system, and weaponry?"

In a few E-min due to sending the message through two micro hyperknots, Admiral Cranshaw responded, "The data from Hari reports the craft to be oval in shape, about 150 m in diameter with transparent exterior protected by some reinforced polymer. It is unclear what drive engine they have, but the fact the spacecraft entered Human Space indicates the aliens can navigate through hyperknots. We know the firepower they have consists of at least thermal missiles. Whether they have fusion weapons or laser power is unverified at the moment."

"Thank you, Steely. General Presoff, do we have an assessment of whether this spacecraft is a scout that is the vanguard for a general invasion of the Federation, or is this a lone renegade intruder?"

"Negative, Madam High Governor; we have no intel on the intent of this attack. Attempts on Hari to communicate with the vessel have, at this date, received no response."

"Is this non-response intentional, or do they not understand any of our coded messages? If we are dealing with an alien civilization, communication would be a major obstacle towards resolving this crisis."

"I suggest," Reg Jones interjected, "the only response they would understand is a swift and decisive retaliation on our part. That is a universal language."

"I am afraid Colonel Jones is correct," said Steely Cranshaw. "With your order, Madam High Governor, I will send a fleet to the Trappist-1 system immediately."

"That is well and good, Admiral. The problem is that at the earliest, the Federation fleet would not get to Hari for at least two E-w. By then, all personnel on Hari may be lost. Yes, I do order an attack force to launch from Teegarden, but we have here on Kanva a starbomber that I plan to command for a more immediate response to this aggression."

"Alyie, we do not know the extent of the alien ship's capability. You are putting yourself at risk by going to Hari," was Reg's agitated reply.

"Colonel Jones, it was you who provided the *Argonaut* and 500 Panthera to accompany me to Kanva. Apparently, you were prescient in my need for these. I now see their value. It will be great to be flying a fighting machine again."

"But this not is the role of a High Governor. Is it Admiral and General?" implored Reg.

"Well," General Presoff mused, "we have never had a High Governor with this flight and military credentials before. I have every confidence in her."

"Thank you, General. I will make my preparations immediately. Let's stay in contact on the encrypted emergency hyperlink line. That is all for now."

Chapter 18.

It took longer than Captain Kap?stk anticipated to secure the perimeter to affect a safe landing on this planet. The creatures' defense system was particularly strong, and Alux had all but exhausted its energy supply with the expenditure needed to maintain the cloak of invisibility and the continued barrage of thermal bombs. Now that it was on the ground, the re-energizing would commence, and the ship would be ready for the return trip in one B?alamian-day (i.e., two E-d.). The creatures fought hard, but they were eventfully subdued by the superior forces of the B?alamians. Not being able to communicate with them and demand surrender, the Captain had no choice but to direct the landing brigade to kill all who resisted. Kap?stk was, although, able to take a few prisoners. Both the living and the dead would have value. The living would teach the victors how to operate the alien technology. In addition, bringing back a sample of these creatures to the home world would be quite a whisker in the Captain's muzzle. The dead, on the other paw, could be put to immediate use on this planet.

The Control Room at the creatures' main building was a mass of panels and instruments. The systems' were run by haptic control. However, Captain Kap?stk and their crew's digits were too thick to effectively operate the consoles. Under threats, a creature was forced to operate the system. (The Captain marveled how undernourished these creatures were. Their paucity of fur and hair made them look hideous in his iridescent eyes. It proved how weak they were in that they had to don outer coverings to increase their body heat.)

Once the communication system was operational, an image of a vehicle appeared on the screen. It was on a trajectory to Hari having emanated from the largest planet in the Trappist-1 system. Not being able to communicate with the alien at the controls, Captain Kap?stk ordered the Alien Technology Officer, Hi?nigh Selm?dom, to identify the type of spaceship.

"Captain," said Officer Selm?dom, "from its size and speed, it looks like a spacecarrier. We are not detecting the signature of any offensive weapons on the ship. This must be a regular delivery of supplies here. Communication Officer Ben?ib Akal?een has intercepted communications that appear to be in the form of a general mobilization of the creatures from their home world. They are launching a fleet of ships against us. But by the Navigation Officer's reckoning, it won't arrive in this system for seven B?alam-days."

"Good," answered Captain Kap?stk, licking its whiskers, "we will be well gone by then." "Force our creature to grant clearance for this carrier to land. We will provide a special welcome for it."

Chapter 19.

Alyie mustered the Panthera and the flight personnel with urgency for their mission to Hari. She assumed command of the *Argonaut* but charged her now co-commander Captain Lenora Honeywine with a special assignment. Co-opting a stratagem used by the Achaeans, a group of renegade robots who masqueraded a ship that sent Alyie and her Valkyrie Squadron on a futile chase, she had Honeywine run an illusion program that projected the image of the regularly scheduled spacecarrier, *Proteus*.

"Lenora," Alyie declared while slyly winking to Honeywine, "these aliens are using transparency to avoid detection; we will one up them with a complete transformation."

"Yes, Madam High Governor," Honeywine chirped. Lenora was a brown-haired young woman with light brown eyes and a light complexion who stood in awe to be having a conversation with her hero.

As *Proteus* (a.k.a, *Argonaut*) entered Harian space, X-ray imaging found the alien spacecraft sheltered in a hanger powering up as if readying to launch. As the starbomber went into orbit around the planet, infrared scans aboard the ship detected about 80 life forms on the surface. The different thermal signatures indicated there were only 20 humans among that group. The thought of the deaths of almost all the personnel at the Institute bore into Alyie. She realized she was dealing with a rapacious foe. All her discipline and control were needed to not order a general strike to the surface. Alyie reminded herself that the prime mission was now to save the lives of those remaining.

Proteus was granted clearance to land by the Flight Controller on Hari, just as Alyie had anticipated. "Cargo" craft carrying Panthera in space suits were launched from the mother ship, ostensibly to deliver their shipments to the surface. Alyie and Honeywine stayed on board

Argonaut to coordinate the assault with the starbomber's crew, with about half of the remaining Panthera serving as a reserve force. As Alyie donned her Admiral's double-breasted white jacket, she attached the rose that Tessa had given her in the top left buttonhole. She hoped it would provide the good fortune the Federation needed.

The landing crafts, as planned, spread out over a wide area to confuse the aliens who were lying in wait to ambush them. Coming under heavy assault from the Panthera, the aliens suffered heavy losses. Their weapons, which appeared to be some type of metal projectiles, were inferior to the laser power of the Panthera. Alyie, using the scans, noticed a contingent of aliens, with some humans making their way through the fighting ostensibly to board the spacecraft, which was powering up just outside the hanger. Directing Honeywine to launch a directed laser missile to disable the spacecraft's propulsion system, Alyie watched as the beam found its target and destroyed the motor drive in a plume of fire and smoke. The aliens realizing all hope of escape was forlorn, surrendered to the oncoming Panthera, releasing their hostages.

"Prepare to take us down, Captain," Alyie said as she completed her preparation for disembarking by inserting a super thin oxygen tube in her nostrils. "It's time to meet these intruders and make them accountable to the Sidereal Federation's Justice."

Chapter 20.

All had gone badly for Kap?stk and the crew of the Alux. As the Captain stood surrounded by the alien soldiers, it pondered on the events that had just transpired, thinking to itself, "Who could have anticipated these creatures would be so deceptive in sending a ship from nearby with such firepower? Our scanners were only focusing on the attack force from their home world and didn't recognize the bomber for what it was. Also, the weapons used by their landing party were much more lethal and effective than ours. Previous civilizations the B?alamian Empire had encountered had inferior weapons and strategies of warfare. There is much we can learn from these aliens, but I doubt I will be able to share it with my pride."

As Kap?stk was standing under guard with its comrades, the huge starbomber landed some distance away. Emerging from the spacecraft was a phalanx of land vehicles that approached in double file. Once in the enclosed oxygenated hanger, the entourage parked and, getting out of the front vehicle, was a tall creature flanked by a shorter one. The two were almost immediately surrounded by armed aliens in uniforms similar to the attack force. The two in front were dressed differently, indicating they were of higher status than their escort. In fact, from the deference given the tall personage and the material attached to its outer garment and cap, Captain Kap?stk presumed this was a high exalted peerage, perhaps even the Emperor of these creatures.

When the leader and their comrade came up close to inspect the prisoners, the Captain noticed their faces were more refined and delicate than the soldiers. "Is this species bigended? That is disgusting. Or are those of the ruling class fairer than the rest? " Kap?stk suppressed the thought that these two creatures would be tenderer than the others, although the Captain's mouth involuntarily started salivating.

As the leader of the creatures was approaching Kap?stk, the Captain realized its moment of destiny was at hand. Hidden under a metal coating inside the right sleeve that enabled it to be undetected was a dagger. It was to be used by a B?alamian soldier to take their own life if necessary. Kap?stk saw the opportunity to literally strike a blow for B?alam by assassinating this alien Emperor when it came near. The Captain would surely be killed, but Kap?stk would be honored among the ancient high halls of the B?alamian Pride as the Wielder of Death.

Just as the leader came close to Kap?stk, the Captain spied a red flower attached to the left side of the creature's upper body. "That's where I will strike," Kap?stk resolved; "it must be the location of a vital organ."

Releasing the catch on the dagger, it slid smoothly into the Captain's right paw, and with a fluid motion, Kap?stk struck hard into the left chest of the creature. However, instead of penetrating through the uniform, the blade was stopped and began vibrating violently as if electrified. Kap?stk was writhing in utter pain as if shocked. Unable to release the blade, the Captain fell to the floor, yowling. After the blade stopped vibrating, the Emperor of the creatures pressed its tall black right boot on top of Kap?stk's hand, kicked away the blade with its left boot, and motioned to the guardsmen standing around to pull the Captain up. Kap?stk, somewhat dazed, could only think, "What manner of creature is this?"

Chapter 21.

After conducting a memorial service on Hari for the slain institute personnel and the fallen Panthera, Alyie made preparations to return to Kanva. (A mystery occurred in that there were eight staff members unaccountable.) Ten alien prisoners remained, and they were placed in custody aboard *Argonaut*. Two of their kind who were killed in the attack were placed in a cryogenic chamber for dissection on Kanva. Alyie requested two exobiologists - Dr. Samuel Kohn and Dr. Mai Pham, and a linguist, Dr. Devon Ricketts, to accompany the return group. Contacting Jaynis and Revita, Alyie reassured them about her well-being. Video calling to Tessa and Tomus, Alyie expressed regret to her daughter. "Your beautiful flower was destroyed when I met the aliens, but it brought me luck because, see, Mommy is fine."

After contacting her parents, Tanita Starbright and Alex Striker, Alyie made arrangements with the Queen and Kanva's Parliament Leader to prepare a holding facility for the captured aliens. As the *Alux* was being hoisted inside the huge cargo bay of *Argonaut*, she reconnected with Jaynis and requested that he form a team to learn all about the ship.

"I want to know what type of materials this craft is comprised of, its propulsion system, and its weaponry. Most important is finding where it came from and the purpose of the aliens' mission. I realize, Jaynis, we are on vacation, and spacecraft engineering and orbital dynamics are not your specialties, but frankly, I feel a sense of urgency, and you are the best I know."

"Alyie, in the past, you have dumped a lot of problems on me, and now you are giving me perhaps the most significant issue of our time – learning about the technology of a hostile alien civilization. Do you think, for a moment, I would pass up this opportunity? I am your man, Madam High Governor."

"Thank you, Jaynis."

"I will be contacting people on Mars, Earth, and other places to form my investigative team. I hope to have them ready as soon as you get that ship over here. But you owe me the completion of our date. I am holding you to that."

"Yes, Dr. Bottombrook, it's a deal."

When her call to Kanva was completed, Alyie hyperlinked a video message to Steely Cranshaw, Ivan Presoff, and Reg Jones. Briefing them on the successful retaking of Hari, Alyie ordered the fleet to cease and desist its mission. However, she maintained the Code V emergency alert and issued an order to the Regional Governors in all sectors of Human Space to prepare for a possible invasion.

"I need to determine as soon as possible," Alyie declared, "whether this alien incursion is a forerunner of a general invasion into our space or just a random raid."

"I agree, Madam High Governor," said General Presoff, "but wouldn't it be safer for you to return to Teegarden and command the operations from Headquarters?"

"I don't have the luxury of time to make that trip," countered Alyie. "I am setting up Kanva as The Center of Alien Studies and am bringing with me some experts to learn all we can about these invaders. From the captured vessel, Dr. Jaynis Bottombrook is forming a group to learn all about their technology. My highest priority and biggest challenge is to be able to communicate with them. We need to know their intent."

"Excuse me, Madam High Governor," interrupted Reg, "the fact that they attacked a benign Federation outpost without warning, killing almost all personnel, and then tried to assassinate you tells us all we need to know about their intent."

"Point made, Colonel. Then the value of communicating would be to impress upon them the repercussions their home world would incur from us if they persist in this aggressive behavior."

“I still believe, Madam High Governor, the best way to show our resolve is a swift and decisive retaliation. But I agree that the more we know about them, the better we can affect this response.”

After dismissing Cranshaw, Presoff, and Jones, Alyie contacted Reg on a private hyperlink line.

“I am sorry, Madam High Governor,” said a contrite Reg, “if I stepped out of line going against your more diplomatic approach to these aliens. I hope I didn’t cause any embarrassment.”

“No worry, Reg. I always want your frank assessment, and there is a part of me that agrees with your sentiments. The reason I called you was to thank you for forcing me to wear that protective mesh. I am alive today because of your insistence.”

“You are welcome, Madam High Governor.”

“How many times now is it that you have come to my aid? Colonel Jones, you are really indispensable.”

“I am just serving my commanding officer, Madame. By the way, how did things go with you and your family on the trip to Kanva?”

“It was a little frosty at first, but I think Jaynis and I were starting to warm up, and then, of course, this crisis hit.”

“I want the best for you, whatever that may be. I am content in any role I serve for you.”

“You are something, Reg. You make things really difficult. I wish you weren’t so fine.”

Chapter 22.

First Communication Officer Ben?ib Akal?een was just finishing the midday meal in the compound where the prisoners were put upon arrival on the primary planet in this star system. On the trip from the spaceport after the bomber landed at the facility, the B?alamian observed how much forestry and plains there were on this land mass. Akal?een mused how much the Hallowed Ancestors would have enjoyed roaming through the thickets and valleys. How much game was in these woods?

Each prisoner was put in their own enclosure, and food was delivered three times per cycle. From trial and error, the captors concluded that Akal?een preferred a diet of fish and meat with some dairy products. Lately, the Communication Officer has been visited by a group of aliens who seemed interested in learning about the B?alamian Empire. In particular, one of them, a tall skinny creature, has been trying to communicate by finding words and sounds that have the same meaning in both languages. For instance, the creature showed Akal?een a round object, wrote the characters "b","a","l","l" on a thin slate rock, and signaled for the B?alamian to write and sound the equivalent. Going through interminable sets of objects, gestures and emotions, a vocabulary list was made for both languages. From the skinny creature continually pressing the digits on its right appendage to its head, Akal?een surmised the information gathered in these sessions was apparently being input into some internal device.

Akal?een at first had some qualms about cooperating with the requests of the captors but rationalized that being a communication officer necessitated striving to understand an adversary's language. This knowledge might be valuable if Akal?een were either freed or returned to his pride on B?alam. When the creature, whose name Akal?een learned was "Devon Ricketts", next came to visit, there was a change in the learning experience. Writing in its language, "I am now rolling a ball towards the wall," and then doing it, Devon asked Ben?ib to sound and write the equivalent expression in Rhilun. Several of these demonstrated actions were then used by the program in Devon's head to establish the grammatical equivalent of both languages.

Chapter 23.

Alyie set up a Command and Control Center at the palace on Kanva. Revita placed her Royal Staff at Alyie's disposal. The Queen's Lady-in-waiting, Rebina Ectot, was designated as a liaison between the High Governor and Kanva's Parliament. The current Leader, Mr. Epson Alandro, and former head, Mr. Dev Gotsan, who Alyie had worked with before, were kept abreast of all developments affecting Kanva. Captain Lenora Honeywine assumed the duties of Alyie's Adjunct. Her primary responsibilities were arranging and coordinating remote meetings with the High Governor's Board and disseminating the information the task forces Alyie had formed concerning the aliens.

Despite the frenetic pace of events on Kanva, Alyie was able to spend time with Tessa and Tomus, which was a great tonic for the intense pressure she was under. It was soothing to have meals with them, assist in their learning, and project holographic stories to them at bedtime. There was a birthday event for the twins, but it was much more muted than originally planned. Jaynis was also in hyper-mode with his task force's analysis of the alien spacecraft. He had set up a hyperlinked team with two astronautic engineers, an orbital dynamic mathematician, and an astronomer.

Jaynis had informed Alyie that while she was on the mission to Hari, Revi had made the proposal to Jaynis about donating his sperm to perpetuate the Jadeite line. Asking for Alyie's input, "Are you ok with this? It is not that I am having sex with her."

"Revi is my best friend and anything I can do, I will to help. Besides, I can provide a great reference for your progenitive power. It is fine with me. Go for it if you wish."

In five E-d. after her return to Kanva, Alyie began receiving reports from the respective investigative task forces. Jaynis came rushing into Alyie's office one day just after she had finished washing and was getting dressed. She had put on her tan breeches and brown boots, but her upper half was only adorned with a tank top undershirt. He couldn't resist saying, "I like your new look. I assume it's in deference to the balmy Kanvaian climate."

"Ok, smarty shoes, what do you have to tell me?"

"We have not completed the assessment of the spaceship's technology, but the big news from decoding the ship's database and log records using triangulation is we know its point of origin. Give my team's gratitude to Dr. Ricketts for developing that handy translator. The home world is in the constellation Leo. It is orbiting the K-class star Sagarmatha. On our star maps, the planet the ship was launched from is Laligurans. It is Earth-like, with a solid core and a gaseous atmosphere. The translation names their planet BʔalamB. Note my attempt at sounding glottal stops."

"Jaynis, why haven't we noticed this system before and sent drones to scan for habitat sustainability?"

"It is 172 l-yr. away, which, my dear, is currently beyond our or, should I say, your limit of Human Space."

"To travel that far to the Trappist-1 system must mean they have mastered faster than light speed."

"There is no evidence of that aboard the ship. Their propulsion systems appear to be some liquid hydrogen, gaseous chemical mix, which would make them slow and inefficient by our standards. We have found stasis units. Our feeling right now is that the majority of the crew was placed in hibernation, with a few members perhaps taking shifts to operate the power systems and navigate the several hyper knots they would need to jump through to reach here in a reasonable lifespan. It must have taken several E-yr. for them to get to Hari. The fact that these aliens undertook this journey implies either a great curiosity or a desperate need to explore uncharted space."

"Keep working, Jaynis and give my thanks to your team for identifying B?alam," Alyie concluded, making her first attempt at the planet's pronunciation.

By the time Alyie met with the medical team which was performing the autopsies on the slain aliens, she was fully dressed in her double-breasted jacket and beige cravat pinned with the High Governor insignia.

"Please come in Drs. Sonoa and Benfield. What have you discovered about the aliens' physiology?"

Dr. Ruha Sonoa, a slightly built dark skinned middle-aged woman, answered, "High Governor, they are definitely mammalian with a robust circulatory system, and their pulmonary system indicates an oxygen-breathing lung capacity. The appendages are more claw-like than ours, but they have full movement in their four limbs and digits but regularly stand upright on their two hind legs. There is also a vestigial tailbone. The abundance of hair suggests an ancestor more fur-like than ours, and their whiskers perhaps lead back to an evolutionary trail that is feline."

"Feline?" Alyie mused. "This would be a fascinating area for academic study if it weren't so imperative we learn how to defeat them."

"Do they have augmented intelligence?"

"No, Madam High Governor," said Dr. Sonoa. "There was no evidence of implants or any non-organic devices in their bodies."

Dr. Adam Benfield spoke somewhat hesitantly in a high-pitched nasal voice. "Madam High Governor, there are two anomalies we found in the dissections. One is peculiar while the other is ominous."

"Go ahead, Doctor; I need to know everything."

"We have found these aliens are bisexual, the correct term being hermaphroditic, said Dr. Benfield while nervously running his long fingers through his silver-white hair. "It is unclear if they reproduce asexually through their own gametes or with the sperm of one organism fertilizing the eggs of another."

"This gets curiouser and curiouser", Alyie said as she leaned forward on her elbows, pressing her hands to her head. "You did mention something ominous, Doctor. Surely it was not their sexual behavior."

No, Madam High Governor," replied Dr. Benfield swallowing hard. "In cutting through their midsection, inside what serves as intestines, we noticed undigested human remains. Apparently, these beings feed on creatures such as ourselves."

Suddenly Alyie was taken aback by the fearful realization of the evidence she was provided. "These aliens are the descendants of predators. They are not looking for new colonies to habitat but new prey to hunt and feed off. And we have become their game."

Chapter 24.

Captain Kap?stk realized something was different this day as four of the black-suited guardsmen approached his compartment. As they spread apart, the tall creature who the Captain viewed as the Emperor approached, dressed in an unadorned white uniform and black boots. The creature signaled to the guardsmen to release the lock on Kap?stk's quarters. Stepping inside, the creature gestured to Kap?stk to sit at one side of a small table that the Captain used to eat its meals while the Emperor sat opposite after taking gloves off and putting them inside its head cap. The fact that this creature was willing to meet one-on-one even after Kap?stk had tried to kill it impressed the B?alamian.

Tapping its head with its hand, the creature spoke, and miraculously, Kap?stk could understand mostly what it was saying.

"I am High Governor of the Sidereal Federation. My name is ????? Starstriker. Please state your name and rank."

Kap?stk was dumbstruck by the creature's facility with Rhilun. The Captain couldn't quite understand its full name, but there was no doubt this "Starstriker" had supernatural ability.

"I am Captain Thu?glr Kap?stk, of the B?alamian Starfleet, Commander of the Alux."

"Good, Captain Kap?stk. You and your remaining crew are our captives. We are members of the ????? Race. You have violated our covenant by invading our space and taking ????? life."

"I don't understand what you call yourselves?" Kap?stk asked.

"Our species is 'h','u','m','a','n' – human", the Emperor said, slowly pronouncing the word.

"Humans?" Kap?stk thought. "They certainly are an ugly race", although the Captain could see why this Starstriker was the leader.

This one was more attractive than the others, with bright red hair and sparkling green eyes, not that dissimilar to the Captain's own orbs.

"Captain, please, state your reasons for your action against our people," the Emperor requested.

Surmising that a show of strength was the only response this human would respect, Kap?stk proclaimed, "The Alux is charged with finding new hunting grounds for my people. We have exhausted the food supply in the space we have already conquered. Your people are clever, but none can withstand the onslaught of the B?alamian Pride. We will overwhelm you, eat your flesh and suck your bones."

"That declaration leaves little space for negotiation, but Captain, do you speak for all your people? I am the leader of mine and the only one who can authorize war on another civilization."

Taking somewhat aback by Starstriker's insinuation of overstepping rank, Kap?stk stumbled on a response. "Roi Sing?a XV, Blessed be the Mane, rules the Pride of B?alam. No, I do not speak for the High Exalted One, but faithfully carry out the commands I have been given."

"Thu?glr, if I may refer to your first name, and mine is spelled 'A', 'l', 'y', 'i', 'e'. I am sure you are a dutiful soldier to your leader and people. But before our two empires come to war, which no doubt would mean terrible losses on both sides, don't you think the proper protocols are for me as leader of the humans to speak to Roi Sing?a XV, the High Exalted Leader of B?alam?"

This one was much too clever and perhaps devious, but Kap?stk had to accept that this "Alyie's" logic was correct in that, as a captain, Thu?glr had no authority to declare war.

"If I can return to my spaceship, I will instruct you how to reach the High Exalted One."

"Thank you, Thu?glr."

Chapter 25.

Before committing to contact the leader of the aliens, Alyie video hyperlinked Admiral Cranshaw, General Presoff, and Reg Jones to get their input, she also included Deputy High Governor Ronica Richlieu in this confab.

"The risk, as I see it, is that if these aliens are as predatory as they appear, the more they know about us, the more they can exploit our weaknesses," said a concerned Ivor Presoff.

"Alyie, what do you hope to gain by this meeting?" asked Steely Cranshaw.

"At the minimum, I hope to impress on His/Her Exalted One that the human race is not to be viewed as livestock. Any attempt to devour us will be met with the strongest opposition," replied Alyie. "A secondary goal," she continued, "is to gain some insight into their civilization that we can use to our advantage."

"I can tell you are resolved to make this exchange," chimed in Reg, "and it may help us, but under no circumstances allow that Captain to speak to its leader. From what I understand, the translation program is still rudimentary and may not pick up subtle messages these B?alamians, or whatever they call themselves, give to each other."

"Clear and understood, Colonel Jones," said Alyie, trying to suppress a smile.

On the day of the meeting, Captain Kap?stk was shepherded by four Panthera to the *Alux,* where Alyie was waiting. She was outfitted in full regalia as befitting the leader of one species about to meet the counterpart of another: snow white jacket with medals of distinction, white breeches, tall black boots, cap, and gloves. Around Alyie's waist, she wore a belt that carried a holster with her laser pistol and a scabbard with the bejeweled hilt of the Jadeite sword she had borrowed from Queen Revita jutting out.

They all entered the spaceship, and, under orders from Alyie, Kap?stk instructed one of the guards to turn on the auxiliary power to operate the communication system. Once it was functional, the Captain gave the code for the guard to initiate the contact. After some time (in that the message was going through several micro hyperknots) video reception was reached at the B?alamian home world. Given a nod from Kap?stk, Alyie tapped the translator in her implant and began,

"I am Starstriker, the High Governor of the Sidereal Federation, and I am aboard your spaceship *Alux* with Captain Kap?stk, who is my prisoner, as you can see. I need to speak to Roi Sing?a XV on matters urgent to both our civilizations."

After a wait of some 30 E-min, the screen became clear, and a creature appeared. It was majestic and monstrous at the same time. The face was a mix of yellow-white fur and hair, gleaming eyes, and long sharp teeth. Its large head was topped by a gigantic crown that appeared to be of gold. What could be judged a smile appeared, and the alien spoke these words that were translated in Alyie's brain:

"I am Roi Sing?a XV, the High Exalted One of the Pride of B?alam; it is good to meet my prey before the hunt begins."

Alyie replied, "I must warn you that my civilization does not consider itself fodder for your people. You have killed a number of my kind, and for that, those responsible must be held accountable. But if you persist in your advance, you and all your B?alamians will suffer sorely. We, humans, have the mind, the means, and the will to utterly destroy you."

After another delay came the response from Sing?a. "You cannot stand in the way of our destiny. Menhit, the great God in the sky, has granted us supreme dominion. Your people are next in our conquest of space. Sleep easy, as we will make good use of your flesh and blood."

"We have the Captain and several crew members here. What about their fate?" Alyie asked.

"Do with them what you want. Their flesh will taste good. Captain Kap?stk has served B?alam well. We have all the coordinates we need to find you and your civilization. Sufficient honors will go to the Captain's pride."

"The fact that we are having this conversation proves we are more advanced than you. You must know you cannot win against us," Alyie tried to appeal to Sing?a's survival instincts.

After an agonizing wait, Alyie heard these words from the High Exalted One:

"Your weapons may be better than ours, and you may kill more than we will, but we will outlast you. Death for the cause is glorious. Soon a multitude of ships will descend on your space. Nothing can stop us; Menhit wills it. I hope we meet in combat. It will be an honor to fight you. If I am victorious, I will pay my respects by feeding on your heart and brain. Goodbye, Starstriker."With that, the transmission ceased.

Chapter 26.

Alyie Starstriker sat brooding in her office suite on Teegarden. In the two E-m. that followed her transmission to B?alam, much had happened. She and her entourage had left Kanva, bringing the prisoners and the *Alux* with them. Before going on to Federation Headquarters, the *Argonaut* touched down on Earth, where Alyie, Jaynis, Tessa, and Tomus made the quick trip to their home in Montpellier. Jaynis had agreed to take a leave of absence from the Université de Paris to join the Federation's Scientific Research Institute to continue the work of his team on understanding the B?alamian's technology and track their movements. However, he stated he needed time to close down his research operations on Earth. Alyie, who believed she couldn't delay her return to the Teegarden, only stayed overnight. It was decided that Jaynis and the twins, sometime later, were to join her. Once reunited, they would all reside in the mansion near the Headquarters building. Somewhat disconcerting to Alyie was that upon arrival in Montpellier, Alison Berenwithy's smiling face was there to greet them. The enthusiastic response from Jaynis, Tessa, and Tomus only exacerbated Alyie's realization that she had more work to do to make her family life whole again.

On Teegarden, there was a whirlwind of meetings, planning sessions, and battle simulations. Alyie had formed a War Council consisting of Reg Jones, Steely Cranshaw, Ivor Presoff, Ronica Richlieu, Mr. Alphonse Cache, head of Cache Space Enterprises, and Ms. Daiyu Fong, The Common Exchange's Director. She was about to have a short briefing with this group prior to a scheduled address Alyie would deliver Federation-wide about the crisis.

Alyie was under pressure unknown to her before. Her eating habits were erratic, and sleep was a rare commodity. To keep herself balanced and focused on the task ahead, she began an intense physical fitness program directed by an autobot trainer, Atlas-5. She also, somewhat reluctantly, agreed to have appointments with the autobot system psycho therapist ELIZA-XXV. These sessions were arranged by Hilda Barton under the highest priority security and took place in Alyie's office. At first, Alyie had difficulty being forthright with ELIZA, but she had become gradually more comfortable sharing her behavior and anxieties with the autobot. The fact that Alyie's parents were on Teegarden gave her some support, but she dearly missed the company of Jaynis and her children. What could be tying up Jaynis so long, she wondered?

Joining the War Council members for this briefing was Dr. Sonja Benita, Director of the Federation's Scientific Institute. She was there to provide the latest findings from Jaynis' Alien Technology Task Force.

When they were all seated around the High Governor's desk, Alyie began, saying, "I have called this meeting to make sure I have the up-to-date information on the aliens' activities before I address the public. I think we need to hear from Dr. Benita first. Please, Doctor, if you will."

"Madam High Governor, Dr. Bottombrook's group has just detected a mass of objects leaving B?alam. From spectroscopic analysis, these objects are of metallic construction, and from the formation, they are assuming, Dr. Bottombrook's Task Force has come to the conclusion this is an invasion force."

"Do we have an estimate," Admiral Cranshaw inquired, "of the number of these spacecraft?"

"It is a little difficult to distinguish distinct objects at this stage," answered Dr. Benita, "but early estimates are about 5000."

That number produced a shock among those gathered in the High Governor's suite, particularly Alyie, who felt her old chest wound throbbing.

“Do we know,” queried Reg Jones, “what the ETA into Human Space is?”

“Given the optimal speed of the ship we captured, and the difficulty of traversing the multiple hyperknots with such a large fleet, the Task Force estimates this force will reach the outskirts of Human Space in a minimum of four E-yr.”

“Four years.” Alyie intoned. “That, then, is our target date. We need to work backwards from what we determine is the ultimate deterrence to this invasion and lay out the means, method, and timeline to achieve that end. I am invoking Emergency Protocol I, which invests in the High Governor’s complete authority over all sectors in the Federation. The Act of Self-Determination will be suspended during this crisis, and I will charge the Regional Governors to have direct jurisdiction over their respective planets. You will be my arms and legs as we mobilize. This is an existential event in the history of the Sidereal Federation. Together we must prevail.”

After some chatter among those present, Alyie preceded to get ready for her address to the citizens of the Federation. She was dressed in a non-military suit consisting of a forest green jacket, skirt, ivory-colored scoop top with a string of pearls around her neck, and black heels. Only the insignia on her lapel indicated her high office. Alyie wanted to project an image of calm self-confidence. The last thing her address wanted to create was a general panic. Speaking with poise and clarity, Alyie began,

“Fellow citizens of the Sidereal Federation, I come before you bringing urgent news but also assurance that the threat you have been hearing about will be met with might and vigor by the forces of the Federation. I have formed a War Council consisting of General Presoff, Admiral Cranshaw, Colonel Jones, Deputy Governor Richlieu, Mr. Alphonse Cache and Ms. Daiyu Fong who will assist me in preparing the necessary deterrence. We have deduced that the invaders will reach our space in no less than four E-yr. For that period,

I have invoked Emergency Protocol I, that will enable the war Council to better coordinate the defense systems throughout all sectors. During this time span, we aim to cause the least disruption to your lives. Citizens not involved in the defense measures should be able to go about their normal routine without undue hardship. We will provide an indestructible shield that will keep you safe from these invaders. I will devote my whole essence to preserving the spirit, ideals, and fabric of the Sidereal Federation. Thank you for your support."

When the address was concluded, Alyie received compliments from all those in attendance, even though she felt a sense of doom in making promises that she was unsure she could keep. Even under the best circumstances, the Federation would never be the same again. As the last of her War Council filed out, Alyie felt inside her a panic attack beginning and called out,

"Colonel Jones, a word, please."

As soon as Reg Jones closed the outer door to the High Governor's suite, Alyie activated the security shield, shutting out the responsibilities of her office, her family turmoil, and the threat of annihilation from the B?alamians, at least for a little while.

(To be continued in Part II.)

The Reign of Alyie Starstriker – Part II.

There is only one God and his name is Death, and there is only thing we say to Death: “Not today”

Syrio Forel, The First Sword of Bravos*

Chapter 1.

As a cooling airflow filtered into the diffuse-lit living room, Tanita Starbright and granddaughter Tessa Starbrook were playing a holographic war game projected on a large oaken table. They were using their implants to direct their respective forces in a simulated space battle. Tessa, nearly 13 E-yr. old, had a recent growth spurt, and was now at least a head taller than her grandmother. She was a student of military warfare and posed a serious challenge to Tanita's vast tactical experience.

As the battle was reaching a climax, Alyie Starstriker quietly entered the room. Silently dismissing her Elite Guards who had escorted her into the High Governor's mansion on planet Teegarden, Alyie watched in contentment in the background as the avatars for her mother and daughter battled for supremacy. From their proximity, it was easy to see the resemblance between Alyie and Tessa. They both were tall with fair skin, freckles, glistening green eyes, and reddish hair that were all obviously passed on from Alex Striker, Tessa's granddad, and Alyie's father (although Tessa's locks were lighter in color than her mother's). Tanita, on the other hand, was shorter, with dark hair that had gone primarily gray. Tessa was in a body-fitting red jumpsuit with soft black boots, while Tanita was outfitted in a flowing apricot dress and elevated brown sandals.

Taking off her gloves, Alyie unbuttoned one side of her white double-breasted jacket and removed her crossover burgundy tie with the High Governor insignia. She then unfastened her belt carrying her laser pistol holster, placing it and the weapon on a nearby desk. Dressed in her trademark white breeches and high black boots, Alyie strove over to the cabinet, where she reached in to take out a bottle of genuine Irish whiskey. Pouring the elixir into a goblet, she walked toward the women still enjoined in battle. Seeing her daughter, Tanita stopped the game and turned to hug and kiss Alyie. Noting the glass with the liquor in Alyie's hand, Tanita scowled, saying, "Should you be drinking that?"

"Oh, Mother, this is just to relax me. The day was a dust devil of activities. I need to have something to help me let down."

"Aren't Tessa and me enough balm for you?"

"Of course you are, but this is a form of lubrication. Let me be."

"Hi, Mom," said Tessa, who had just noticed her mother's presence. "How are the preparations for the invasion coming?"

"It is an ongoing process. We had several meetings on deployment, strategy, and new weaponry. The new class of star bombers, Zeta class, is about to be rolled out. We're having a formal ceremony in three E-wk."

"Yes, Reg, I mean, Colonel Jones told me," said Tessa. "He said I could attend."

"He did? Well, if it is ok with the Director of Sidereal Federation Security, it is fine with me. Who am I, the High Governor, to step over his authority?"

"Also, Mom, Colonel Jones has indicated I can intern in his office in logistical planning for the Panthera. It would be a great experience. I would be following in Grandad's footsteps."

"I am sure you can be an asset to Colonel Jones and his staff. Go for it."

As Tessa excused herself to give instructions to Taki-4 to prepare dinner, Tanita and Alyie stayed in the living room to converse.

"Speaking of your children," Tanita interjected. "What is going on with Tomus? I haven't been able to reach him in two E-d."

"No worry, Mother," said Alyie. "I spoke to Jaynis today, and he said Tomus is now doing his own internship as Special Assistant to

Alison Bernewithy at the Ministry of Arts and Culture. He is interested in art history and wants to learn as much as he can about Earth's past and culture. I think he has found his niche."

"It's good for me to hear that you and Jaynis still communicate."

"Of course, we share two beautiful children; they form a bond that will never break."

"But it is so sad that you have become so separated; you here with Reg, and he on Earth with Alison. I never thought you and Jaynis would grow apart."

"Mother, life has a way of imposing change on us. I still love Jaynis as much as ever, and I feel deep inside him he feels the same about me. But let's not dwell on the things that divide us. Tonight we three will have a wonderful dinner together. I understand Tessa has Taki making Shepherd's Pie."

Chapter 2.

As the light filters were just beginning to open up, Alyie came downstairs into the dining room, where her mother was already at the table. Alyie had not finished dressing for the day but had put on a white button-down shirt, gotten into her tan breeches, and pulled on brown riding boots. She and Reg Jones had promised each other that they would have a twice weekly ride across Teegarden before their work day began at Headquarters. The invigorating early "morning" gallop helped Alyie clear her mind of the vicissitudes she would face later.

"Good morning, Mother. I hope you slept well last night. Do you know when my father is getting back from Mars?"

"I video-linked him after dinner last evening. He is scheduled to leave today, Mars time. He should be here on Teegarden in four E-d."

"Good. The War Council needs any updated predictions my father's Telemetry Group has on which hyperknots the aliens would use to jump out of in their first strike against us. Jaynis' B?alamian Deterrent Group will then use analytics to propose the most effective counterforce stratagem against the invaders."

While the two women were softly conversing, an autobot entered and served a breakfast consisting of synthetic eggs, assorted "meats," and biscuits. Juice, coffee, and tea were also available on the sideboard.

Tanita asked Alyie, "Dear, what kind of a day does it look like for you?"

"The same as any other, Mother - incredibly busy. I have a War Council meeting as soon as I get to HQ. Alphonse Cache is briefing us on the status of the Luna manufacturing facilities' efforts in building the new Sidereal Fleet ships. Daiyu Fong is providing a report on the Common Exchange financing of the military campaign.

She will also update us on the distribution of Exchange funds to ensure food and supplies are allocated for an extended siege across the Federation if necessary. Later, Reg and I are meeting with Steely Cranshaw, Ivor Presoff, and other commanders in my Strategic Task Force to devise deployments to repel and defeat the invaders."

"Alyie, with all that, is happening, do you regret running again for the High Governorship? Perhaps if you hadn't, your personal life wouldn't be so disruptive."

"Yes, Mother. We have been through all that. It would be a nice step back from all this turmoil, but retiring to a quiet life on Earth will not make the threat of the B?alamians disappear. I was at the beginning, and I saw the face and the ferocity of their leader Roi Sing?a. Their threat of our annihilation is real, and I think it would be inexcusable to back away at this stage. You and Father have instilled that sense of duty. I am the product of your value system."

"Don't misunderstand me, Alyie. I do not want myself, my husband, daughter and grandchildren to be on these beasts' menus. I pray every day that we can prevail against these monsters. But I am very concerned about the toll this crisis is taking on you."

"If the citizens of the Federation wanted me to step down, they could have made that happen in the election three E-yr. ago. I need not remind you that I won another term by an unprecedented margin of victory, Mother."

"It would have been shocking if you were voted out in the middle of a crisis. It was great enough a worry for the citizens to ignore any questions about your personal conduct."

"Mother," said Alyie with her voice rising, "despite your thoughts to the contrary, I have never shirked my responsibilities nor let emotions dictate my decisions. I have a very supportive team that has helped me in my tenure as High Governor, including Reg, Jaynis, and both you and Father. As long as you are with me, I think I can succeed."

"I am sorry to speak out of turn, but, my dear, I am so worried about you."

As Tanita got up to give Alyie a teary hug, Reg entered the dining area, dressed in dark blue military riding attire. Seeing the two women in an emotional state, he started to back up until Tanita spied him and beckoned him to come forward.

"Oh, Reg, don't mind us; we are just having a mother-daughter moment. Please join us for breakfast."

"Thank you, Commander Starbright. How are you doing, Madame?"

As Reg approached the table, he stopped by where Alyie was sitting and placed his left hand gently on her shoulder. She, without turning around, brought her right hand over his.

"If you must be formal, I am fine, Colonel Jones. We were just discussing our mutual favorite subject and her propensity for working till exhaustion. I am relying on you, Colonel, to keep her in check."

"Understood, Commander. I will treat this as Job 1."

"And while we are on the subject, I want you to promise me with all that you consider sacred that under no circumstances will you allow her to lead the assault fleet against the aliens. We both know well enough that is just what she intends to do."

"Madam Commander, on the soul of my ancestors, I promise as Federation Director of Security, I will prevent her even if it means disobeying the High Governor's direct order not to do so. So I swear."

"Excuse me," an exasperated Alyie exclaimed. "Doesn't the person in question get to say something in her defense?"

"No!" both Tanita and Reg chimed in chorus.

Chapter 3.

After finishing breakfast, Alyie and Reg excused themselves to go down to the new stable facility that Alyie had ordered constructed. Alyie had completed her outfit by buttoning her shirt and fastening a yellow tie around her collar. She donned a sleeveless bright yellow button waistcoat and, with Reg's assistance, put on a brown and orange stitched tweed riding jacket. Before leaving, Alyie had turned to her mother saying, "At a suitable time, could you gently wake up and remind Tessa that if she wants to begin working for Colonel Jones, she should be fed, dressed, and ready. With my schedule today, I won't have much time to wait for her when we return."

Alyie and Reg walked side-by-side down the path behind the mansion to the barn. Thorin, Alyie's black 13 E-yr.-old stallion greeted the couple's arrival with a loud neigh. (Alyie had him transported from her chateau in Montpellier on Earth when it became clear that she would be staying primarily on Teegarden.) Reg's mount, Atlantis, was a tall 12 E-yr.-old brown gelding. Before mounting, they fastened spurs to their boots, put on their riding helmets, and Alyie stuck a riding crop in her right boot.

Once out of the stable area, the horses ran freely over what served as the surrounding meadow. Reg was the superior rider and an expert horseman, but Alyie had a good seat in the saddle. Not taking any chances, Reg had a squad of Panthera hovering above the couple in airjets providing the necessary security for the High Governor and her Director of Security.

After a strong gallop across the plain, the pair took rest at a familiar leafy-shaded area that was adjacent to an artificial pond. Dismounting, Reg and Alyie brought their horses down to the water's edge. They walked arm-in-arm, taking their helmets off, enjoying each other's company. As Alyie peered into the still dark water, she saw the image of a mature couple; she was nearly 44 E-yr., and Reg was in his mid-50s. How she had changed from her early days as a Valkyrie Commander some 23 E-yr. ago. There was much that Alyie remembered as good, but also some regret, especially in her unfaithfulness to Jaynis. However, she could not deny her feelings for Reg and the comfort she found in his strong, steady presence.

As she continued staring into the water, Alyie grew pensive, and suddenly her face and Reg's dissolved, and the image of Roi Sing?a XV peered, leering back at her. Fearing she was about to be swallowed in his gigantic mouth, Alyie jerked her head up, gasping.

Reg grabbed hold of her arm, saying, "What's wrong? What did you see?"

"I saw Sing?a's face," was Alyie's frightened response. "It was there in the water mocking me and my efforts to ward off the attack."

"Alyie, calm down. This is just your imagination. The invaders are still 1-yr. away."

"I keep thinking I am not preparing enough for their invasion. What am I missing?"

"Rest assured; we will prevail against them."

"We think we have the superior weapons, but what if we are underestimating their arsenal?"

"Alyie, that is what the Deterrent Group is trying to ascertain. Once they come up with an assessment, the Strategic Task Force can optimize the counterforce."

"But, Reg, there are many examples in human military history of inferior armies overcoming superior opponents. These B?alamians are ferocious and fearless."

"Yes, Alyie, these aliens may be all that, but we have three distinct advantages. One, we have more advanced technology and better training, as evident by how easily the Panthera dispatched them at the

Soffen Institute. Two, we are fighting for our very survival while they are just looking for new hunting grounds. And, three, but most importantly, we have a great leader who will outthink and outmaneuver Sing?a the 20th or whatever his or her number is."

The compliment seemed to steady her, and drying her eyes, Alyie said, "It is XV, by the way. Well, if I am going to outthink and outmaneuver this Alpha feline, I will especially need your assistance."

"My mother always said, 'Reggie, when a superior asks for your help, don't question, just say, how can I serve?"

As her anxiety attack subsided, Alyie oriented herself so she faced directly at Reg with her booted legs so close together that her spurs clicked. She gazed into Reg's soft brown eyes and put her gloved hands on his cheeks. Coming together, they kissed passionately as he enfolded his strong arms under her jacket to encircle her body.

Chapter 4.

When their sojourn by the pond was over, Reg helped Alyie button her waistcoat and refastened her tie in place. As they remounted for the journey back to the High Governor's mansion, Reg thought it was the time to bring an issue forth that concerned him.

"I know we agreed not to talk business during our morning rides, but looking at your schedule, I am not sure I will have your undivided attention anymore this day. Can we discuss something?"

"I think, Colonel Jones, you have earned the right to break protocol today. Ask away."

"What is the fate of the alien prisoners we brought here from Kanva? My staff is receiving a good deal of pressure to try them for murder and then have them executed."

"I don't want that to happen, Reg. They are to be considered prisoners of war. I know their treatment of the staff members of the Exobiological Institute on Hari was horrendous, but I want to demonstrate that humans, unlike B?alamians, do not eat their enemies."

"Alyie, have you forgotten human history? Many ancient Earth cultures engaged in feasting on the remains of their opponents. In some societies, this was a way of honoring a fallen foe."

"Colonel Jones, are you suggesting we revive that practice? I was unaware of your dissatisfaction with the food at the Headquarters Commissary."

"Madam High Governor, you know I was not talking about reciprocating their behavior but bringing them to justice for their unwarranted attack on our people."

"No, we are going to hold them here and treat them with respect and dignity. I don't see them as monsters but as a species that is driven by the need to hunt. We need to find a way, while protecting ourselves from their ravenous appetites, to redirect their primal instincts toward less destructive consequences."

"Excuse my impertinence, but, Alyie, you are sounding now more like a behavior modification autobot than the High Governor of the Sidereal Federation."

"I will give you a second pass on being insubordinate to your commanding officer but don't try your luck anymore," was Alyie's playful reply. "To show how serious I am about 'civilizing' the aliens, I am inviting Captain Thu?glr Kap?stk to have a private luncheon meeting in my office tomorrow."

"What! He, she, or it tried to kill you. What are you thinking?"

"No worry, Reg. I will be wearing that nice mesh protection you outfitted me with previously. The more I converse with the Captain, the better I can understand the B?alamian culture. I have to believe it will help in our defense. By the way, the best our translator comes up with for the proper third-person singular pronoun in Rhilun is spelled 'seh.' So in referring to Captain Thu?glr Kap?stk, use seh for he/she, 'shim' for him/her, and 'shis' for his/hers."

"If you insist on being in the same room with this creature, I want the Elite Guards to be there as well."

"No, Reg. I want to put Captain Kap?stk at ease. It will be just the two of us. I would, of course, expect the guards to scan the Captain thoroughly before the meeting, and they are welcome to be stationed outside."

"Rest assured, one of those outside will be me."

"I wouldn't want it any other way," spoke Alyie while extending her left gloved hand across her mount to touch Reg's right hand.

Upon their return to the stable, Reg and Alyie dismounted, and after attending to their horses, they walked back up the path to the mansion. Once inside, Alyie immediately went into her bedroom suite for a quick wash and change. Shortly after, Tanita and Tessa came into the anteroom, greeting Reg. Tessa was dressed in a formal Panthera uniform of a black "leather" double-breasted jacket and blue pants tucked inside tall black boots. She had her black gloves inside a cap that she was holding in her right hand. She looked at the picture of youthful energy and beauty. At seeing Reg, Tessa snapped her boots together and saluted him.

Reg was clearly astonished and impressed at Tessa's appearance and demeanor. Collecting himself, he assumed the role by saying, "At ease, Ms. Starbrook. I see you are ready for your first day. Welcome to the Department of Federation Security."

"Thank you for this opportunity, Colonel Jones."

While Reg, Tanita, and Tessa were chatting, Alyie emerged from her bedroom after replacing the tweed sports jacket and waistcoat with her white double-breasted jacket adorned with medals of her rank and office and fastening her High Governor pin on her yellow tie. Taking close notice of Tessa's appearance, she gave her daughter a warm hug and said, "It looks like we are all ready. Let's get going. See you at the end of the day, Mother."

Chapter 5.

As Alyie anticipated, her work day was a whirlstorm of meetings. At the War Council confab, Alphonse Cache updated members on the pace of the production on Luna of the new Zeta class starbombers. Mr. Cache, Head and Founder of Cache Enterprises, was a white-skinned, slightly built, middle-aged man. The new model spacecraft featured enhanced laser and fusion weaponry in terms of range and firepower. The propulsion system used a more efficient scoop to collect intra-stellar hydrogen fuel. The spacecraft was also larger to house over 1000 troops, more material, and an updated drive system.

"Mr. Cache," said Alyie, "are we on schedule for the delivery of 500 of these spacecraft to the Fleet?"

"Yes, Madam High Governor, the autonomous manufacturing systems are running almost continuously. You should have your new fleet in less than an E-month."

"As they roll off the assembly, I want batches of them flown to Headquarters immediately. The Strategic Task Force needs to deploy them throughout the Federation. We do not know when or where the aliens will strike first, and it is urgent we have our counterforce in place as soon as possible."

"Consider it done, Madam High Governor."

"Ms. Fong, have you secured funds for the acquisition of food and supplies for an extended siege?" Daiyu Fong, a petite pale-skinned young woman dressed in a tailored black pantsuit, was an economist at the Common Exchange.

"Yes, Madam High Governor, the Federation Bank has covered the cost to the Common Exchange with a low-interest mega-loan. We have dispatched space freighters to the food-producing planetary systems to collect and then distribute the rations to the neediest worlds throughout the Federation."

The remainder of the War Council meeting consisted of questions concerning the current location of the aliens' invasion force and what their initial target would be.

Admiral Steely Cranshaw, Head of Space Command, fielded the question of location. "The Telemetry Group has the invasion force splitting into five squadrons of about 1000 ships each and jumping through different hyperknots. It is not clear if they are planning simultaneous attacks in different locations in Human Space or trying to create diversions to confuse our tracking capabilities."

"It is my understanding that Dr. Striker has obtained updated information on their whereabouts that he believed was too sensitive to transmit over hyperlink," continued Admiral Cranshaw. "I believe he is currently returning to HQ from consulting with his colleagues on Mars."

"The Deterrent Group needs this intel immediately," said Ground Forces Commander General Ivor Presoff. "They cannot do their assessment of where the attack will commence with old data. I am disappointed that Dr. Striker does not have confidence in our secure communication system."

"General Presoff, my father is somewhat of a skeptic in terms of trusting Space Command's hyperlink transmission. Let me speak to him over my channel which has additional safeguards. I think I can get him to share what he has learned."

Chapter 6.

After the War Council meeting concluded, Alyie was escorted back to her office suite by Reg, who left her after making sure all was secure and safe. Ordering a quick lunch served by an autobot, Alyie made some calls before convening a meeting of the Strategic Task Force. Her first video hyperlink was to Jaynis, who was in his office at the Université de Paris.

"Hello, Jaynis. I just wanted to check in with you. I hope all is well."

Jaynis, who now adopted a civil but unemotional tone with Alyie, responded by saying, "My group is growing impatient with your father's Telemetry team. When are we getting the update?"

"I am doing fine, too; thank you, Jaynis. I am aware of the problem, and I will contact Father today and see if he is comfortable enough to share the data with me. As soon as I obtain it, I will send it over to my High Governor high-security link to you and General Presoff. I am sorry for the delay."

"Tessa informs me she is interning for Jones. Did you encourage that?"

"No, this is Tessa's own doing. I am way too occupied to make such arrangements."

"Is this a case of like mother, like daughter? Look at the role model you serve for her."

"Jaynis, you are free to think whatever you want of my relationship with Reg, but I assure you, as her mother and on Reg's honor, this internship will be purely professional."

"See to it that it stays that way."

"On the subject of our children, my mother has been unable to reach Tomus lately. What is he up to, and how can I contact him?"

"Oh, he has been literally buried in the archives of the Louvre. The transmission to his implant is probably poor down in the bowels of the museum. Alison gave him a project to unearth all art acquisitions housed in the complex that deal with aliens. She has the idea of making this into a special exhibition. Sort of putting an artistic perspective on how humans perceived their potential annihilators."

"Well, I hope this exhibition is launched as a retrospective after the impending invasion is over."

"Let me send you the archive hyperlink contact information. You should be able to reach him."

"Despite your work with the Deterrent Group, have you been able to find time to continue your research at the Université?"

"I am glad you asked. We have been able to isolate mini-black holes, and using gravitational fields, we have found they can be manipulated to effectively close hyperknots. It is based on a paper by Hans Bruckner, a 22nd E-c. Physicist. This might come in handy if you want to block some of the routes these aliens are using to come to and from Human Space."

"This sounds promising, Jaynis. Please continue your work and let me know when you think these black holes can be utilized. Stay well."

As soon as the call to Jaynis was concluded, Alyie made the hyperlink contact to the Louvre's Archives. When the transmission was completed, she saw the bright face of Alison Bernewithy instead of her son's.

"Hello, Alyie, or excuse me, Madam High Governor," exclaimed a somewhat surprised Alison. She was dressed in a fine-tailored gray suit consisting of a jacket and skirt. Underneath the jacket was a pale pink blouse with a large bow around the collar. Her now reddish hair was pulled back and held in place with a black headband.

"Hello, Alison. How are you? I am trying to contact Tomus. Jaynis provided me with this link. Is he around to speak?"

"Yes, he is. I must say that Tomus is such a natural for historical scholarship. He and I are having a grand time going through the Archives."

Alison turned and waived. Presently Tomus appeared. He was tall and lanky with his father's sandy-colored hair and blue eyes, but he had Alyie's freckles.

"Hi, Mom. I am having a great time looking through these files. There is so much human history in the art the Louvre has collected over the E-c."

"I am happy you are enjoying yourself. Looking at humanity's past is fundamental to understanding our nature. I wanted to see you and also ask you to contact Grandmother Tanita. She really wants to see and speak to you."

"I will, Mom. How are things coming with the defense preparations? When do you think the attack will begin?"

"In answer to your first question, things are going well. The second question does not have a firm answer as of yet, but we at Headquarters are continually monitoring the alien fleet's trajectory."

"I hope I get to see the aliens in person. I really didn't get the chance when they were on Kanva. You always kept the prisoners away from us."

"My job, Tomus, is that no one in the Federation gets to see the aliens up close. By the way, where are you staying during your internship? You don't commute back in forth to Montpellier, do you?

“No, Dad and I stay with Alison in her apartment in Paris. She and Dad are planning to buy a pied-à-terre which will provide more room for all of us. Dad mentioned that you and he had lived in one years ago.”

“Yes, we did,” Alyie sullenly replied.

Chapter 7.

Somewhat disturbed after her conversation with Tomus, Alyie tried to be sanguine about these developments and focus on the tasks in front of her. She used her High Governor hyperlink channel to reach her father. Making audio contact, Alyie greeted Dr. Striker, hailing, "Hello, Father, this is Alyie. Where are you?"

"Alyie, is that you? How were you able to reach me? I have been blocking messages from Presoff and your husband."

"Being High Governor comes with some perks, including being able to get through to an obstinate dad. Are you in a spacecraft?"

"Yes, I am a passenger in the space cruiser *Eurydice*. We're heading towards Luna."

"Aren't you supposed to be traveling back to Teegarden?"

"I will, but I have requested a quick stop at TDRSS. We have been receiving some spurious signals that I need to confirm are not blips."

"So you are going to the Tracking and Data Relay Spacecraft System Center on Luna. Do you think the signals are from the aliens?"

"That's what I need to confirm. This process is pretty challenging, but we're making slow progress on identifying their trajectories."

"Father, slow is not a word I like to hear in these circumstances. Jaynis' group needs your best assessment to develop algorithms for optimizing our defense strategy, and my Strategic Task Force needs Jaynis' analysis to deploy the Fleet and Ground Forces. Please give me what you currently know."

"Alyie, it is not that simple. Let me walk you through the problem. An alien fleet was launched from their home world Bʔalam in the constellation Leo about four E-yr. ago. The planet's star, Sagarmatha, is 172 l-yr. away from Teegarden. The 5000-strong armada traveled about two E-m. to their system's nearest hyperknot. Jumping out, the fleet divided into roughly five equal parts and traveled in different directions. Despite their transparent hulls, the Telemetry Group was able to track their paths due to the sheer number of spacecraft. Now it gets more complicated. Each of the five subgroups traveled on average about 1½ E-yr. to a different hyperknot. From relativistic celestial mechanics, we know each of these jumps have multiple exits. Given an average of three alternative routes in each hyperknot, this leads to 15 potential exit paths. Our task is to ascertain where each subgroup came out. We identified three of the subgroups' directions. That leaves two uncertain (with three possible routes in each) making a total of nine possible paths in all."

"So we need to set our forces outside of these nine hyperknots to meet the aliens when they emerge."

"Oh, Alyie, if only it were that simple."

"What's the complication, Father?"

"After coming out of these secondary hyperknots, we project this invasion force traveled on the average another 2 E-yr. to a tertiary set of hyperknots. Again using orbital dynamics, we estimate three routes per jump, which makes a total of 27 possible exit points."

"But Father, there have to be less than 27 hyperknots in Human Space."

"Alyie, there are exactly 12. So one or more is where they will emerge."

"But we knew that number before any of your group started tracking the invaders. What help has your telemetry given us?"

"That's why I am going to Luna. We noted some abnormal transmissions. Suppose we can identify them to be from the advancing fleet that might narrow the possible emergent points significantly. I am hoping to reduce the number to perhaps 18."

“But Father, that does not help us concentrate our forces. With that high a number, we would still need to defend all hyperknot exits to provide maximum protection.”

“Talk to Jaynis; maybe his Deterrent Group can use probabilities to gauge where the most likely invasion points will be. It might also help to have some insight into B?alamian battle tactics.”

“Maybe I can aid in that second matter. But send me what you have, and I will forward this to Jaynis and present what you give me at my Strategic Task Force meeting later today. Be safe, Dad. I love you.”

“Love you back, Alyie.”

Chapter 8.

At the Strategic Task Force Meeting, Reg Jones, Steely Cranshaw, and Ivor Presoff were joined by Captain Olga Fercouska, Valkyrie Squad Commander, and Rear Admiral Lenora Honeywine. (This year's Deputy High Governor, Fernando Alvarez, also sat in attendance.) Alyie shared the information Alex Striker had communicated to her. All were frustrated by the ambiguity in identifying the emergent points for the invasion.

"Dr. Striker suggests it would help narrow his Telemetry Group's search efforts if we try to predict based on military science where the aliens would most likely first strike," Alyie offered as a constructive exercise for those present.

"If their goal is to pick off as many of our people as possible, I would think they would choose isolated targets such as Kanva and Ergo, which are at the fringes of the Federation," argued General Presoff.

"On the other hand, if they want to achieve a decisive stroke, they may come right at us here on Teegarden," voiced Rear Admiral Honeywine.

"My feeling is," said Olga, "they want us to divide our forces by launching their assault from multiple jump points. Why else would they have split up their armada?"

"That split has already been accomplished as currently the Fleet is spread out over the entire Federation in trying to provide some form of deterrent in every home world," Admiral Cranshaw retorted.

"I need to gain more insight into the B?alamian psyche. Do they practice stealth and deception, or is that below their sense of pride and honor? Without that knowledge, we are just assuming they would act as we would," surmised Alyie.

After the meeting, Alyie returned to her office and continued working until the end of the day. She was picked up by Reg, who brought Tessa went him. The young girl was bubbling with enthusiasm over her first experience at the Security Office.

"Mom, Colonel Jones introduced me to everyone in the Department. They all were so friendly and helpful to me. I attended the morning staff meeting, received a tour of the facilities, and was assigned a workstation. Reg's, I mean Colonel Jones' Adjunct, Lieutenant Brianna Crosby, tasked me to begin cataloging the existing security systems currently in place across Federation sectors. I am scheduled to work two days per E-w. I promise it won't interfere with my studies. I can't believe how nice people are to me."

Alyie did not give voice to her thought that being the daughter of the High Governor might have a lot to do with Tessa having received such a positive welcome at work. Instead, she said, "I am happy things went so well. I am sure you will be an asset to the Department. Remember, you have a family tradition of service to uphold."

"I promise I will, Mom."

"Colonel Jones, I am holding you responsible for seeing that she works hard and is treated like any other staff member in your department," said Alyie in a mock serious tone of voice.

"Yes, Madam High Governor, you can count on that," was Reg's exaggerated reply.

Taking the short ride to the High Governor's mansion, Alyie, Tessa, and Reg were greeted upon arrival by Tanita, who directed Taki-4 to set the table and prepare a dinner for four.

Chapter 9.

Captain Thu?glr Kap?stk was ushered into the High Governor's Office Suite. The Captain and shis nine fellow B?alamian prisoners were housed on Teegarden in a spacious compound that provided them with amenities such as outdoor roaming areas, comfortable living quarters and physical facilities. They were free to converse and interact with each other, and although hyperlink communication was limited to local sites, they could use it for entertainment. Kap?stk had never been to the High Governor's office but had interacted with her several times as Alyie often came to visit the prisoners. She still commanded to shim an aura of superiority from when the Captain was unsuccessful in shis assassination attempt. As befitting an audience with the leader of the Humans, Kap?stk dressed in a formal uniform (made to order by shis captors) of light green single button jacket, green pants with a yellow stripe (that matched shis fur) indicating the Captain's rank. As typical for a B?alamian, Kap?stk was unshod.

As Thu?glr entered the inner office, Alyie was seated behind her large wooden desk. She stood up to come and greet shim, dressed in a bright blue pants suit with a white satin blouse underneath the jacket. A black crossover tie was around her collar that complemented her tall black heels. Her red hair was let down and let to fall freely on her shoulders.

Alyie extended her right hand with the High Governor's signet ring to the Captain, activating the translator in her implant, saying, "Greetings Thu?glr, I hope you are well, and you have all the amenities you need. I am also so glad you agreed to join me for lunch in my office."

The Captain, not quite prepared for this graciousness on the part of who seh considered the enemy, took Alyie's hand in shis left paw and held it, softy.

Using the limited facility of the enemy's language that Thu?glr had acquired, seh responded, "Thank you for inviting me to share food."

As she invited the Captain to sit opposite her at the table in her office, Alyie couldn't help noticing shis halting movements. "Thu?glr are you unwell? I can have my personal physician examine you."

"It is just my age, High Governor. I am approaching the end of my time. We B?alamians do not have long lives like you Humans do. It is the way of life."

Alyie then used her implant to order autobots to cart in the meal. The Captain was served fresh raw meat from cattle that were raised specifically for the B?alamian diet, while Alyie had a fresh fruit and vegetable salad. A pitcher of water was available in addition to a bottle of Gamay that was synthetically fermented at the Headquarters Commissary. Pouring the wine into each of the two goblets at the table, Alyie raised her glass, offering a toast, "Here's to a good meal together for your health and the hope for better relations between our two species."

After picking up Alyie's cue that seh should reciprocate, Thu?glr said haltingly while raising shis goblet, "Good hunting and clean kills to you, High Governor."

"Please call me Alyie. Tell me about your family Thu?glr. I have mentioned that I have twin children; they are named Tessa and Tomus, that are now 13 E-yr. old. My daughter Tessa is in Teagarden and would like to meet you while Tomus resides on Earth. Both are the product of the union between myself and Jaynis Bottombrook."

Neither the children's names nor "Jaynis Bottombrook" translated into Rhilun, the language of the B?alamians, so Alyie had to spell each out slowly for the Captain to understand.

"I have a large Pride," was Thu?glr's response. "We couple with many different partners in that our sex is changeable, so I am a member of many different families. But since I am the oldest and have the highest rank, I am the head of all of them."

"It is not unusual for us also to have more than one sexual partner, but usually, the family unit is pretty fixed. You must miss them terribly. Let's toast to all of our families well-being. Cheers! This is what we say before we drink."

"'Cheers' back at you, Alyie," said Thu?glr after draining his glass. "Yes, it has been difficult, but I am serving my people even in confinement."

"I tried to negotiate your release, but your ruler was pretty adamant against you returning."

"Yes, I am aware of that," said a subdued Captain. "My people don't exchange prisoners in the belief that captivity contaminates them."

"Do you feel we have contaminated you?" asked Alyie after refilling both their wine glasses.

"You have treated us well, much better than how we deal with the prisoners we take. That has changed my opinion of you, and one might say that it has made me 'soft' towards your species. You know I really like your wine. Please, Alyie, some more."

"What you say is 'soft', I would claim, is having a better understanding. I have likewise formed a high opinion of your culture, admiring your discipline and sense of duty and honor."

"There have been many times I wished you would execute me so I can return to the ancient halls of the Pride, but I find I enjoy living here and conversing with you."

"Thu?glr, I do as well. Can't we be an example for the rest of our two races? Does there have to be this war?"

"Menhit, the great God in the sky has willed it, so has Roi Sing?a XV decreed," exclaimed the Captain raising shis glass high. "We B?alamians have been given dominion over all other species in the sky. They are all for us to use as we see fit."

"But our civilization does not wish to serve that purpose to you. Can't we find a way to satisfy your need and preserve ourselves? You know we will resist you with all our might. Most of your people will die who participate in this invasion."

"I don't want to see our two peoples destroyed. I can see you are a noble race," Thu?glr admitted.

"From what we know, your people are on the verge of attacking us. I don't want you to betray your allegiance to all you value, but is there any way you know that we can avoid a mutually destructive war?" Alyie motioned to pour more wine into Thu?glr's goblet.

"Yes, please. War is inevitable; it is in our blood."

"Perhaps we can achieve some limited action that would lead to a truce. If we knew the location that your forces would attack first, we could fortify that area and achieve a stalemate."

"Alyie, I will not tell you that even if I knew the location, but I can inform you that your actions in preparing for this assault have been monitored from our home world."

"How is this possible, Thu?glr?" said a wide-eyed and suddenly sober Alyie.

"Our First Communication Officer, Ben?ib Akal?een, has been able to use your network to tap into your Space Command transmissions across the Federation. Seh has managed to find a way to send it to B?alam. I am sure the oncoming fleet has this information."

"Thank you for sharing this with me, Thu?glr. Rest assured. I will be discreet in using it. Please have more food, and let's uphold the tradition Humans have to always finish a bottle of wine when dining," said Alyie, who was hard-pressed to resist the immediate impulse to warn everybody of the security breach.

Chapter 10.

As soon as the guards escorted Thu?glr Kap?stk out of the High Governor's Suite, Alyie used her secure hyperlink channel to contact Reg.

"I've just learned that the B?alamian armada has had relayed to them the communications from Space Command to our forces throughout the Federation. They are aware of our defense preparations and countermeasures."

"They can't have received any of the exchanges between the special task forces you set up nor the proceedings of the War Council," countered Reg. "I have made sure all such communications were under the highest security protection."

"But knowing our fleet movements and troop deployments has to give them an advantage," said Alyie.

"Let's put our heads together, Alyie. If we were listening to all these messages from Space Command, how could we use them tactically? How would this help determine where to strike first?"

"Where the Federation is most vulnerable and where an attack would have the most impact, I guess," Alyie replied.

"And that is?" Reg wondered.

"Where the new starbombers are on the ground. On Luna," Alyie said with a frightful conviction.

"Then we need to get them in space as soon as possible."

Alyie immediately contacted Alphonse Cache and ordered him to bring as many Zeta class starbombers as flyable to Teegarden. Then she convened one E-hr. later an emergency meeting of the War Council.

At the hastily arranged assemblage, Alyie updated the War Council members on what she had learned from Captain Kap?stk and her fears about Luna.

As she was speaking, an Emergency Code V signal began ringing in everyone's implants. The message from the Communications Center announced,

"A large constellation of unregistered spacecraft have emerged from the Sol System Alpha hyperknot and are headed in the direction of the Earth-Luna system. Their ETA is estimated to be 3 E-d."

All Alyie could think of was her father, Jaynis and Tomus.

Chapter 11.

Amid the frenzy of activity on Teegarden to mount a counteroffensive to the B?alamian assault, Alyie requisitioned her old fighter craft *Grane* from her time as Valkyrie Commander. It had been upgraded now to a Pegasus-20 model. She and her old Valkyrie teammates, Olga Fercouska, Hollie Pederson, Zarah Khoury, and Wang Li would make up, in their respective fighters, the advance wing of the attack force. They will be followed by squadrons consisting of starbombers and stardestroyers. At the optimum, the Federation fleet would reach the Earth-Luna system in four E-d., just after the estimated assault by the invaders. Alyie prayed she would not be too late to avert complete destruction. There would be some resistance by the ground forces, but it was doubtful if it could hold out for long.

In the Cielo Club adjacent to the hanger that housed the fighters, Alyie, along with her teammates, was finishing suiting up for the mission. Alyie chose to wear a version of her old Valkyrie uniform consisting of a tight-fitting dark blue flight suit, highly polished tall black boots, and gloves. Over her shoulders was draped her Admiral's jacket with medals and the High Governor's insignia. Instead of the usual banter among Valkyrie, the mood was decidedly somber as the women realized the gravity of this mission. Using her implants to coordinate with the response team, Alyie was in constant communication with Admiral Cranshaw, General Presoff, and the wing commanders, directing this mission.

While Alyie was talking, Reg walked in and requested everyone else present to leave.

"I promised your mother to prevent you from doing what you're about to do, but I don't see how I can stop you, nor do I disagree with your actions."

"Thanks for being so understanding. You know I have to do this."

"Yes, I do. It is the soldier's way."

"I have issued an Executive Order giving you complete authority for directing the Federation's attack in my absence. If I do perish, Fernando Alvarez will be my successor. I am sure he will need your help desperately."

"Stay alive for my sake and the Federation's."

Reg then came forward, embraced Alyie, and the two kissed long and hard.

"Please take care of my mother and Tessa. They will need your support."

"With all my heart. Vaya con dios."

Before getting into *Grane's* cockpit, Alyie hyperlinked Tanita and Tessa.

"Oh, Alyie, I am so worried about your father, Tomus and Jaynis."

"We hope to get there early enough to discourage this attack," Alyie assured her mother, although she didn't really believe that.

"Go get them, Mom. It is so cool you are a Valkyrie again. I hope I can be one someday. Bye."

Chapter 12.

As the Federation fleet went into the hyperknot in the Teegarden system, Alyie braced for what she would learn on the other side. Her primary responsibility was to protect the manufacturing facilities on Luna, but she kept thinking about Earth and its inhabitants, including her husband and son. When *Grane* emerged out of the hyperjump with the other Valkyrie, Alyie began receiving distress signals from both Luna and Earth. The communication was garbled, but it was clear both worlds had sighted the incoming invaders. They were entering the Earth-Luna orbit. The three E-d. journey to the Earth-Luna system had put the Federation formation one day behind the aliens' ETA. Alyie hoped the Home World's defenses would hold the B?alamian attack force off long enough for the squadrons from Teegarden to intervene.

After numerous exchanges with the Strategic Task Force on Teegarden, other wing commanders, and fellow Valkyrie, Alyie tried to reach her family members who were in harm's way.

"Father, are you on Luna?"

"Yes, Alyie, we have been advised to go into this bunker. The puzzle is solved. We now know where they are attacking first."

"Stay safe. I'm on my way. I love you."

Alyie was unable to reach Tomus but after some difficulty, made contact with Jaynis.

"Where are you, Jaynis?"

"I am at the Université. I can't leave. Everyone has been forced to seek shelter in the building. I can't get in touch with Tomus. He must be deep below in the sublevels of the Louvre. I have contacted Alison, and she has left the Ministry to find him."

“I am leading a space force to intercept the invaders. We are coming. We will stop them.”

“Alyie, don’t put yourself at risk. The Federation can’t afford to lose you.”

“If I don’t put myself at risk for my family, then what good am I to the Federation?”

“I will let you know if I find Tomus.”

“Good-bye. I know you don’t believe me, but I love you.”

Chapter 13.

As the Federation forces were approaching the Earth-Luna system, they sighted a constellation of fleeing new Zeta class starbombers under attack from a group of B?alamian spacecraft. Alyie ordered two wing commanders to break away from the formation and give support to the constellation as she and the rest of the response force began to descend into a high Lunar orbit. Alyie and her Valkyrie then used an Oberth maneuver to accelerate downward. As they were approaching the surface, scans detected the ominous sight of mushroom clouds, a clear sign that the aliens were using fusion bombs. It was evident that the force shields protecting the manufacturing facilities had been breached, and considerable destruction was now evident.

Alyie attempted to contact the Federation Lunar Command Base but received no response. Fixing in her sights one of the B?alamian spacecraft responsible for the bombing, Alyie messaged the other Valkyrie fighters saying, Enemy at 60° with its force field down. Ladies, let's go get it."

"Yes, Captain. I mean Admiral; I mean High Governor or whatever you want to be called," stammered Olga.

"Alyie will be fine," was the High Governor's response.

The five Valkyrie fighters formed a pentagon around the invading bomber and, once in range, commenced firing with their laser cannons. The bomber tried to respond but was overwhelmed by the multiplicity of attacks on its outer hull. Unable to get its shield back up before its drive and life support systems were compromised, the ship was quickly engulfed in fire from the combined Valkyrie assault. Leaving the vessel internally smoldering, the Valkyrie turned their attention to another B?alamian spacecraft bombarding the surface. The other members of the Federation fleet engaged in battle with a force of B?alamians who had been stationed in orbit at Lagrangian Point L2 on Luna's far side, 1.5 million km from Earth. The

Federation stardestroyers had a tactical advantage in speed and firepower over the larger and slower alien spacecraft. The fighting was protracted and fierce, but the Federation eventually held sway.

As Alyie and her group were destroying the last of the five spacecraft that assaulted the manufacturing plants, she received a signal in her Space Command implant,

"Emergency, Emergency, Earth under attack!! Federation Regional Headquarters in Brussels and the surrounding area was destroyed by heavy thermite bomb assault. Aliens landing in Paris and taking hostages. Please assist."

Alyie felt her heart stop at the terror implied. Collecting her composure, she ordered five starbombers from the squadron loaded with ground troops to land at the Paris airport and rescue all humans. She also commanded the landing force to use lethal force against the enemy.

Chapter 14.

As the Valkyrie Squadron circled around the remains of the manufacturing plants on Luna, their scans noticed something unexpected.

"I am not recording any radioactive daughter products from the explosions," said Hollie Pederson.

"Do you think they used pure fusion bombs?" wondered Wang Li.

"Hold on," announced Zarah Khoury. "I am getting readings indicating positron emission. They must have used antimatter for a trigger."

Alyie, who had not held out hope for her father's survival, was unsure if he and the others could survive the after-effects of matter-antimatter annihilation events. Nevertheless, she believed her prime efforts should be directed to Earth.

"I am ordering a Search and Rescue Team to land on Luna. You four Valkyrie are to stay in low orbit to guard against any counterattacks."

"Yes, Alyie," saluted Olga. "Where are you going to be?"

"I am going to land on Earth."

"That is a negative, High Governor," shot back Hollie. "Earth has not been secured, and there are hostile forces on the ground." Her objection was echoed by the other Valkyries.

"I am sorry, ladies, but even though I have been out of the Commander's seat for some time, I still outrank you all."

"Alyie, please be safe and don't do what you normally do, namely lead," said a concerned Olga.

"Yes, I promise to follow behind the ground assault troops," said Alyie, not really convinced of her own statement.

Chapter 15.

Landing in heavy fog within a secure perimeter at what was historically Charles de Gaulle Airport, Alyie was met by a convoy headed by Brigadier Commander Kurt Beckwall, a tall slim, but a trimly built white-skinned man in his mid-40s with brown hair and sporting a thin mustache. As the Commander approached, Alyie could see that the air was filled with smoke, with fires smoldering in the distance.

Saluting, Commander Beckwall announced, “Madam High Governor, as you requested, we have an armed convoy ready for your use. However, I have to warn you that the environs are still conflict zones.”

“Understood, Commander. Please give me a quick briefing of the situation.”

“After launching a series of thermite bombs centered on Brussels but spreading to the surrounding terrain, the aliens landed and combed the area for hostages with the apparent intention of bringing them to their spacecraft. We have intercepted many of these cargo trucks and freed as many hostages as possible.”

“Good work, Commander. What is the damage to the Regional Headquarters Building?”

“I am sorry, Madam High Governor; it has been completely demolished. We fear there are no survivors.”

Alyie groaned at the thought of the people at that facility she knew when she was Regional Governor and made a silent prayer that Hector Palmero, her friend, and colleague, was not in the building during the assault.

“What other areas have suffered from this attack?” Alyie asked.

"Paris was also hit very hard. For some reason, the aliens seemed to single it out. Perhaps they thought bombing the city would drive its citizens into their hands."

"Can you be specific, Commander, at what parts of the city were damaged?" said Alyie fearing the worst.

"It was the city's heart. Many of the most visible structures, such as Notre Dame Cathedral, the Louvre, and the Pantheon, were struck."

Trying to maintain control of her emotions, Alyie announced, "Please, let's get going; time is critical."

"What is our destination, Madam High Governor?" asked Commander Beckwall.

"Paris."

Chapter 16.

As the convoy was making its way through the fog and ash southwest to the center of Paris, there was sighted through infrared scopes the signature of B?alamians troops lying in wait to ambush them some 300 m. ahead. Halting the convoy, Commander Beckwall ordered soldiers in two armored trucks carrying surface-to-surface missiles to prepare the weapons. Once loaded and targeted, the guns were fired at the enemy's position. When enough damage had been done, the Commander gave the order for the convoy to proceed. As they advanced, they came under scattered fire from the remaining alien forces. Alyie, who had her protective mesh on and a safety helmet, had grabbed a laser rifle from the supply trunk in her vehicle and used it to pick off a few stray shooters.

As they entered Paris proper, Alyie told Beckwall their first stop would be the Université de Paris on the South Bank of the Seine River in what used to be called the Quartier Latin. Circling around, it was observed that the Université suffered only minor damage. Disembarking from the vehicles, a squad of soldiers dispatched optical nanobots into the buildings and, after waiting a suitable time, observed no alien presence. Commander Beckwall then gave an all-clear signal on the Université's communication system. Soon after, faculty and staff began to exit the buildings. Alyie had her motor car parked outside the Physics building, and when she saw Jaynis emerge, she rushed to him. Their meeting was heartfelt but restrained.

"I am so relieved that you are unharmed," Alyie expressed.

"Let's go find Tomus," was Jaynis' reply.

Jumping back into the lead ground vehicle with Jaynis on board, the convoy crossed over the Seine to go to the Louvre. The museum was heavily damaged, and sections of the complex were on fire. The structures on the other side of the river were utterly demolished. Captain Beckwall ordered infrared scans to detect the presence of life, and there were readings on the south side. However, some of the life forms were not human.

Donning gas masks and protective outerwear, Alyie, Jaynis, and a troop of soldiers carefully entered the building fully armed. Immediately, they were set upon by a group of five enemy snipers, but anticipating the attack with an additional force coming from behind, the Federation forces dispatched the B?alamians.

Combing the building using the detectors, the rescue group found museum staff hidden in offices and galleries. Finally, Alyie and Jaynis made their way down the emergency staircase to the subterranean levels. Through the rubble and smoke, they reached the sublevel, Records, unseen for centuries by the billions of visitors to the Louvre. As they entered the corridor, a loud noise was heard down the end of the hall as if a small explosion had gone off. Removing their masks and rushing to the room where the sound appeared to come from, Alyie and Jaynis found a door broken down and voices, one of whose they recognized to be Tomus'. Slowing entering with Alyie's hand on her rifle's trigger, they found a dead B?alamian soldier with a hole where their face would be on the floor with wisps of smoke drifting out of the cavity. Huddled behind a desk was Alison, held tightly by Tomus. In Alison's hand was what looked like an old 19th E-c. pistol. Her face and tailored dark suit jacket were smeared with red blood and a creamy white colored substance. As she struggled in her high-heeled boots to get up, Tomus still held her tightly with his hands firmly entwined around her waist, and it was evident she was experiencing great pain from a broken left arm.

"Mommy, Daddy, Alison shot that alien right in the face when it was about to grab us. Isn't she terrific?"

Alison, who was about to go into shock, managed to say, "It's not my brain and blood," before she passed out as Jaynis rushed towards her.

Chapter 17.

A few E-d. later an uneasy calm had settled on Earth. The alien attackers had either been killed or driven off the planet. Those that escaped joined their comrades at the L2 point behind Luna as they were trying to hold their position from the continued attacks by the Federation squadrons. However, most of the Fleet's Zeta class starbombers had been destroyed on Luna. Only about 20 were able to fly safely to Teegarden. From the nuclear attacks on the Luna bases, no survivors were found. Once confirmed, Alyie made the agonizing call to inform Tanita that her husband and Alyie's father had died.

Alison had received medical attention for her broken arm and other injuries from autodocs. She was brought by airbus along with Alyie, Jaynis, and Tomus to Montpellier, which had suffered no damage during the alien assault, to recover. Jaynis had notified Alison's parents and one of her sisters of her injury and assured them she would heal completely. Placed in one of the guest rooms on the upper floor, Alison was waited on attentively by Tomus, who was in awe of what she did in the museum.

Alyie too, was very appreciative of Alison's heroic efforts. During one visit during her convalescence, Alyie expressed, "I don't know how I can thank you enough for protecting Tomus."

"I consider him family. I would do anything I could for him. In fact, you all are so important to me. My life has changed so much since I came to work for you."

"Alison, I know we have been distant recently, and it is mostly on my part. I have come to realize that, in many ways, you have stepped into the roles I should have been taking with my family. I did resent you for that, but if you didn't, Jaynis and the children would have suffered. Thank you for all you have done."

“Alyie, there is no one like you. It is so important that you continue to lead the Federation during these horrible times. Anything that I can do to ease your burden, I will. I am here to serve you in any capacity you wish, even if it means stepping aside.”

“No, Alison. I realize, for now, your place is with Jaynis. He is still a part of me, as I hope I am with him. It’s just that I need to devote all my attention to this war. But, no matter what happens, let’s remain friends.”

“That is all I ever wanted.”

Chapter 18.

Alison stayed an E-w. in Montpellier until her arm was mended enough by the auto-surgeon Alyie had dispatched to attend to her. When it was time to return to Paris and the Ministry of Arts and Culture, Alyie had a new suit made for Alison in that the old one would forever carry the memory of the B?alamian Alison had shot. The new outfit was a dark gray with white pinstripes tailored jacket and skirt suit made of fine wool from sheep that Alyie and Jaynis had raised on their property. (Their 3D printer/processor constructed the clothing.) On the day she would return, Alyie entered her room, finding Alison with her suit jacket draped over her shoulder, struggling due to the sling on her left arm to put her boots on. Alyie bent down on one knee in front of her old Adjunct and finished the task.

"I wanted to speak to you before you left. I know there is a lot of work at the Ministry, especially overseeing all that involves the restoration of the Louvre, but I want to discuss with you the Regional Governor's position. I received the news I had feared that Hector Palmero was among those who were in the Regional Headquarters building when it was bombed. He, along with everyone else, didn't survive. It is a terrible loss for the Federation and for me personally in that I looked on him as a surrogate father."

"I am so sorry, Alyie. I knew how much he meant to you. And with your father's death on Luna, that is a lot of grief to handle. How is your mother dealing with it?"

"I speak to her every day, and Reg has been very attentive to her. On hyperlink, she has been remarkably composed. I don't think the loss has fully set in yet. It does help to have a very cheerful person like Tessa around her."

"I know Tomus has been feeling his grandfather's death. The other day when we were looking at the images of the artwork damaged in the attack, he broke down when he saw Rembrandt's 'Self-Portrait of Artist with an Architectural Background'. I think the figure reminded him of Dr. Striker. The only thing I could think to do was hug him," said Alison.

"I am sure that was a comfort," said Alyie. "He thinks so much of you. But, returning to my proposal, I want you to assume the Regional Governor's position. It will be an Interim appointment, but next E-yr., you will be free to run for it, and, if you do, I will provide my full endorsement."

"Alyie, you have caught me completely by surprise. I don't know what to say."

"'Yes,' would be the answer I want. You are the natural choice. Your time working as an Adjunct for both Hector and myself gives you a unique insight into the running of that office. Your leadership in heading the Ministry of Arts and Culture shows your administrative talent. And as the person who has played a role in your career, I see this as your next step. Please say 'yes'."

"As I said, I am here to serve. Of course, Alyie, I will do it. Who will lead the Ministry?"

"I was hoping you would take on both positions in the interim. If that is too much, then appoint a Deputy to run the Ministry. Whatever you do, I will support you. Also, any resources you need to restore the Louvre and build a new Regional Headquarters will be made available."

"Thanks. Let's see how I make out as head of both. Don't worry; I will be contacting you often for help."

"Good. Now, before you leave, I think Jaynis is waiting to speak to you."

Alyie withdrew from Alison's room as she knew Jaynis would be entering soon. She gave them enough time to be together before calling the airjet to take Alison back to Paris.

Chapter 19.

Despite the crisis, Alyie believed it was her place to be on Earth with Jaynis and Tomus. The stay also enabled her to visit the damage done by the attack that took the lives of over 200 thousand and provide some public measure of comfort to the survivors, many mourning the loss of a family member, friend, or colleague.

Turning Montpellier into a Command Center, Alyie was in frequent communication with Reg and the other members of her Strategic Task Force. Due to the responsibilities, he assumed in Alyie's absence, Reg's Adjunct, Lieutenant Crosby, would often take the call. Brianna was a young woman with a creamy caramel-colored complexion, brown curly hair trimmed down to her scalp, and orange-yellow eyes. Her ample figure was clothed in a light brown jacket and skirt uniform that matched very well with her skin color. (Alyie wasn't sure if her coloring was natural or if she used the services of a pigmentologist.) In her nicely pressed kaki buttoned-down shirt fastened securely by a black tie, the Lieutenant was a picture of poise and competence. Despite her professional demeanor, Brianna's devotion and admiration for Reg was obvious.

"Greetings, Lieutenant. I hope this day finds you well. I would like to speak to the Colonel."

"Yes, Madam High Governor," Crosby responded, standing at attention and clapping her stacked heel black boots while saluting.

Reg soon appeared on the video hyperlink saying, "It appears the primary mission of the assault was the destruction of the new starbombers. Once that was accomplished, the aliens made a diversion in attacking Earth. Either it was in response to Regional Headquarters sending a counterforce against them, or they saw an opportunity to create havoc and seize humans."

"Oh, Reg, you were so correct in criticizing me for my coddling of the prisoners. If I hadn't been so liberal, they wouldn't have been able to tap into our communications and discover our starbomber-building activities on Luna. It's time to take your advice and proceed to have them sent to the Ministry of Justice and be prosecuted for crimes against humanity."

"I know this is out of character for me, Alyie, but I will now argue the opposite and recommend leaving them as is. If your friendly alien Captain has not disclosed to its comrades what it told you, we can use the fact that we know they are listening to tell them what we want them to hear, and perhaps trace any message they send to determine the locations of their forces."

"If we do that, be sure to provide extra security for information we don't want to be overheard."

"Will do, Madam High Governor."

In communication with other Strategic Task Force members, Alyie asked, "Are there any reports of other alien attacks in Human Space?"

Admiral Cranshaw reported, "Madam High Governor, there have been incursions in the Tau Ceti system directed toward the planets Ergo and Cognito and in the Gilese 667 Cc system. The B?alamians now seem to be executing their main stratagem in picking off our worlds one by one in the quest for new hunting grounds. We have sent a squadron with one of our new Zeta class starbombers to each system to provide support. They should be able to fend off the invaders."

"What is the status of the situation around L2?" Alyie inquired. "Has the alien formation shown any movement from this equilibrium position?"

"No, Madam High Governor," responded Rear Admiral Lenora Honeywine. "They seemed to be either waiting for us to advance or hoping to receive reinforcements. Federation forces from the rescue mission have surrounded them and are awaiting instructions on whether to attack or hold their positions."

"Do they have any humans on their ships?"

"No, Madam High Governor, our scanners detect no living homo sapiens," said a somber Honeywine.

"That leads to the only conclusion," General Presoff surmised, "that any hostages they brought on their ships have been consumed."

"Do our forces surrounding them have an overwhelming advantage, Admiral Honeywine?" Alyie asked, trying to keep her composure.

"Negative, Madam High Governor," Lenora responded. "Using our simulations, we have a 70% probability of prevailing but not without incurring significant losses."

"Then let's keep the blockade in place for now and wait in see what they will do."

"Understood, Madam High Governor. I will send an additional squadron to increase our advantage around L2 and provide rotating relief for those enforcing the blockade," answered General Presoff.

"My fear, General, is this will develop into a war of attrition in which we react to each move they make. At that rate, it will take all of our lifetimes to successfully prosecute this defense. Ladies and Gentlemen, we need to find a means to decisively defeat these invaders at a minimal loss to ourselves. That is what I charge this Strategic Task Force to determine."

Chapter 20.

A decisive victory did not come forth as the war dragged on for almost two E-yr. The Federation forces, for the most part, were able to resist the attacks from the B?alamian squadrons as they emerged from hyperknots. Eavesdropping discretely into the exchanges between the captured Communication Officer Ben?ib Akal?een and the attackers, the Federation was able to locate the intruders as they hyperjumped into Human Space. Soon after, Federation squadrons that were trolling near the hyper-exits were sent to intercept the aliens before they reached any habitable worlds. The battles in space were costly in personnel and materiel, but the superior speed and firepower of the Federation craft almost always prevailed. Thus, the planets and outposts of the Federation were kept safe, and the citizens could experience some degree of normalcy in their everyday activities.

Alyie had returned to Teegarden. She regretted leaving Jaynis and Tomus on Earth as they had become compatible again, but she was now able to comfort Tanita and be with Tessa. Alison had assumed the dual role of Interim Regional Governor and Director of the Ministry of Arts and Culture. She had purchased a pied-à-terre on Avenue Montaigne close to where the new Regional Office was being built. Tomus often stayed with her when he was interning at the Ministry. Out of deference to Alyie, Jaynis didn't join them in Paris until the High Governor left Earth.

One day while Alyie was conferring with Reg in her office, Alyie's faithful Adjunct, Hilda Barton, buzzed her, announcing that Lieutenant Brianna Crosby was in the front office requesting permission to join Alyie and Reg. After Alyie granted such, the young Lieutenant strove into the room, stood at attention, and saluted both Alyie and Reg. She then informed them that the Sidereal Federation Command Center had just received a message in Rhilun that, when translated, stated, "The High Exalted One of B?alam requests an audience with the 'Ruler of the Humans'".

“Where is the message coming from?” Alyie inquired.

“Madam High Governor,” answered Brianna, “the tracers have detected its origin in the Tau Ceti system.”

“So, Roi Sing?a is in Human Space,” Reg said, recognizing the significance. “To put him/herself in harm’s way must mean he/she is getting frustrated at our success in foiling the grand plan of conquest.”

“This could be the opportunity we have been looking for,” Alyie exclaimed, as her mind was racing through various scenarios.

“Do you want to speak to it?”

“Yes, I do. Lieutenant, see that a message is sent in Rhilun saying that the High Governor of the Humans will conduct a video hyperlink session with the High Exalted One in one E-hr. At this meeting, we can discuss the terms of the B?alamian surrender.”

“Yes, Madam High Governor.”

“Do you know what you are getting into with that beast?” asked Reg.

“Yes, I do. Roi Sing?a is the head of the pride. Well, it’s through shis pride I will get shim to fall.”

Chapter 21.

Alyie dressed in her most formal attire for this video conference. She wanted to project an image of power, confidence, and authority. After putting on her double-breasted white Admiral's jacket with medals and decorations, she tucked her snow white pants into her almost knee-high black polished boots. She pulled her reddish hair back with a barrette and tucked it under her black beret. A cravat to match her hair was placed around her neck pinned by the High Governor's insignia. After putting on white gloves, she put the signet ring signifying her high office on the fourth finger of her right hand. With her laser pistol in her belt hostler and a ceremonial sword (which she had previously given to Reg) sheathed in a scabbard attached by a belt around her waist, she gazed in her full-length 3D mirror. What she saw was a nearly 46-Earth-year-old woman who was the leader of some 110 billion humans about to meet the ruler of a race that wanted to enslave and then devour them.

Reg entered her bedroom and, looking at her up and down, and front and back, pronounced proudly, "You look ready to do business with this monster. I wouldn't want to be on the wrong side of you today."

"My plan is not to scare shim, which I don't think is possible, but to enrage shim so much that Sing?a does something rash that we can use to our advantage."

Alyie stood before the large screen in the Conference Room for the hyperlink video meeting planting her boots firmly. It took a few E-min. for the connection to be made, but eventually, the image of a feline-like creature with a massive head of yellow and white fur appeared. Unlike the previous occasion, Alyie had seen Roi Sing?a on screen; this time seh was not wearing a crown but had some kind of gear that looked like a flight helmet. Tapping her translator in her implant, Alyie announced,

"Greetings Roi Sing?a XV, Exalted One of B?alam from Starstriker, High Governor of the Sidereal Federation. I am happy you have the need to talk to me."

The face on Alyie's screen appeared to make what looked like a smile as it ground its sharp teeth.

After the expected short delay in transmission, Roi Sing?a replied, "I commend you, Starstriker, Ruler of the Humans. Your kind has put up a good fight. You have proven a worthy foe."

"We are more than a worthy foe, Roi Sing?a. Since your initial surprise attack, we have stopped your every advance, and the squadron that assaulted our home world is marooned in space with no hope of rescue. I think it is time for you to sue for peace."

"We will cease our attacks on your worlds if you agree to our terms. We want 120 million of your people to be given to us. You are free to choose offerings from any part of your system. We ask that your selections be young rather than old in age. That is the price you will pay for our peace."

Alyie, who was prepared for an unsatisfactory proposal, was still, nonetheless, stunned by the audacity of this request. "Your price is too high. Our people are not sacrificial. We will not only defend every single human; we will take the fight to you. Prepare your defenses, Sing?a; we are coming to B?alam. I am personally going to lead the attack as a Valkyrie in my fighter *Grane*."

"We have destroyed your new fleet. What will you attack us with?"

"You think that is all we have. As I speak, we are building new bombers to replace those you destroyed. Even with the existing ships, we are defeating you."

Alyie sensed her show of bravado was making an impression on the B?alamian. Seh appeared to be thinking and when seh resumed speaking, said, "Starstriker, I like your ferocity. It is the trait we value above all in B?alam. It would be a great honor to fight you in one-to-one in combat - you in what you call '*Grane*' and me in my great *K?aal Ch?eeta* fighter. The winner gets to determine the conditions of peace."

"As the leader of my people, I would be foolish to risk their safety on my skill at defeating you. No, we will beat you on the battlefield with intelligence, superior weapons, and willpower. It will be a collective effort, not single combat. However, if our paths cross, either in Human or B?alamian Space, I will not hesitate to shoot you out of the sky. I hold you responsible for the death of my father."

"So be it, Starstriker or 'Valkyrie'," answered Roi Sing?a. "I will be looking for you in *Grane.* May the victor feed on the other's heart and brain."

With that, the transmission ceased.

Chapter 22.

Alyie believed her conversation with Roi Sing?a laid the groundwork for a trap to be set. However, there were pieces to be put together before she could implement her plan. Making what was called a family trip to Earth in a spacecruiser, Alyie arranged a meeting with Jaynis at Montpellier to ostensibly discuss selling the chateau.

When she arrived with her Elite Guard escort, Jaynis was already there sitting casually in a loose-fitting sweater and corduroy pants on the sofa near the fireplace in the dark wooded family room. His light hair was now sprinkled with white, but Jaynis still gave the appearance of a young man full of energy and curiosity. After the house was scanned for safety and a security field was put on, Alyie removed her hunter's green jacket and sat on a chair opposite Jaynis crossing her brown booted legs. She began the conversation by saying, "How are you, Alison, and Tomus doing in Paris?"

"Things are well. Tomus is really thriving in his archival work. He really likes delving into the past."

"That is great. I do miss him terribly."

"I miss Tessa. Video conferences are fine, but I want to be in her presence."

"She is becoming an outstanding and beautiful young lady. You should be very proud, Jaynis."

"Just like her mother."

"Oh, Jaynis. How terrible I have been. I wish this had never happened."

"You weren't terrible; the pressures of the job became too great."

"Please forgive me."

"There may be a time for that, but that is not the reason you have come so discreetly to visit, is it?"

"No, it is not Jaynis. I want to know if we can use the mini-black holes you have captured to collapse a hyperknot. I am trying to trap the ruler of B?alam in one."

"How do you intend to entice shim to enter a hyperknot?"

"I am planning to use myself as bait. I have challenged shis pride. Given the opportunity, I am pretty sure seh will try to kill me."

"This is *so* you. It would be my luck; just when it looks like we might be getting back together, you get yourself killed."

"No, I don't want to die. That is precisely why I have come to you. Explain how this black hole you have captured can close down a hyperknot."

"As you know, black holes are regions in space that are the result of gravitationally collapsed matter formed in different ways. The most spectacular mechanism is the death of massive stars (over three times Sol's mass) which produce a contracted spacetime cavity that accretes matter into it. On the opposite end are micro-black holes which are either specks of stellar black holes that have broken away from their parent or leftover material from the primordial event that created the Universe. The meso-type are mini-black holes which we suspect were the micro variety that, over time, accrued enough mass to be larger. The ones we have found and captured are now thousands of kilograms."

"How are you to keep them from absorbing matter?"

"Easy, Alyie. We use a surrounding perfectly balanced induced gravitational field to keep these black holes at a respectable distance. We have isolated three so far, and they are housed on Mars."

“OK. Now the crucial question. You told me once these mini-black holes might serve a useful purpose in thwarting the aliens. How can that be done?”

“I am glad you remembered, Alyie. When the subject came up, I believe we were trying to determine the routes these invaders would take to reach us. I suggested we use mini-black holes to collapse some of the possible hyperknots that would lead to them entering our space. However, I realized that was impractical because that would close off our ability to travel as well.”

“Yes, but if we just close off one hyperknot, we would still have the rest free to use. Tell me how it can be done.”

“The thing to do is transport the mini-hole using a robotic spacecraft up to the chosen hyperknot. Once it is close enough, we would eject it from the gravitational field so that it enters the hyperknot. Once inside, if the black hole has more mass than the knot which we think ours do, it will exert a strong enough gravitational attraction to completely collapse the knot, effectively trapping anything inside.”

“That would be perfect. I now need to let our adversary know my travel plans so seh will pursue me into the knot before the trap is set.”

“So you are entering first and hoping the alien will follow. You realize you run the risk if this works of being trapped inside forever as well.”

“I am relying on you, Jaynis, to keep me safe.”

“Haven’t I always?”

“Of course.”

Chapter 23.

When Alyie returned to Teegarden, she convened a meeting of the Strategic Task Force. Lenora Honeywine provided an update on the B?alamian presence in the Tau Ceti system.

"Madam High Governor, they now have a formation of perhaps 400 spacecraft that have made the hyperjump into that system. Their fleet has not yet made a direct assault on the two inhabited planets: Ergo and Cognito. There have been engagements with our forces patrolling that area, but so far, there has been no major battle. Isn't the ruler of B?alam in command of this formation?"

"Yes, Lenora," said Alyie, "and it is my intent to draw him out of the protection of his fleet."

Alyie then informed the members of the Task Force of her plan to trap Roi Sing?a in a hyperknot, using herself as a decoy. Not surprisingly, objections were raised, most vehemently by Reg Jones.

"This is insane, Alyie; I mean High Governor," said a stammering Reg. "All Rules of Engagement forbid putting the Commander-in-Chief at risk in combat. Don't you realize what a demoralizing effect it would have on our campaign if we lose our leader?"

"That is precisely what I want to happen to the B?alamians if we can dispatch their ruler. This war has gone on too long. This may be the means to end it. My plan is not subject to debate as I have already set it in motion."

"What support do you need to advance this mission, Madam High Governor?" asked Steely Cranshaw.

"Dr. Bottombrook and his colleagues are securing the object to a robotic spacecruiser on the Martian spaceport. I will contact Ergo and Cognito directly, and I plan to visit them ostensibly to inspect the defensive systems and bolster their morale. The hope is that the message will be relayed by our eavesdropping prisoner here on Teegarden to Roi Sing?a and his forces in the Tau Ceti system. It would make the visit seem more legitimate if the fleet could provide my Valkyrie fighter a military escort."

"Will do, Madam High Governor," answered Admiral Cranshaw.

"I will be riding with Olga in her fighter as your wingman," asserted Reg.

Alyie could see from the strength of his remarks this was non-negotiable.

"OK, Colonel. Welcome to my soirée."

Chapter 24.

Due to the intense preparations needed for her trip to the Tau Ceti system, Alyie took up residence again in her office suite, working almost continuously while her daughter and mother remained in the High Governor's mansion. One "morning," Tessa was ushered into the inner office by Hilda Barton. The young girl was not dressed in her Panthera uniform, but instead was clad in the light brown jacket, snow white pants, and brown boots of a Fleet Academy trainee. With her bright green eyes sparkling and her face flush with freckles, she blurted out, "Mom, I have been accepted into the Flight Academy Prep School. My uniform arrived today."

"Congratulations, Tessa. I am so happy for you."

"I am following you and Grandmother. I start next E-m. in New Cedonia. Isn't it great?"

"I have no doubt you will be successful. You have it in your blood."

Tessa pivoted in her boots and, from a plié position, executed a pirouette in front of her mother to show how well the uniform fit on her. When she was finished showing herself off, Tessa's mood became more somber, and she turned to Alyie, saying, "Mom, you should visit Thu?glr. I think seh is dying."

Alyie had been aware that Tessa had been a regular visitor to Captain Kap?stk and the other prisoners. Alyie, who had not been seeing shim that regularly, took heart that her daughter took a special interest in Thu?glr. From her conversations with Tessa, Alyie was aware that Tessa and the Captain played games together, each exposing the other to their own culture's favorites. They also practiced speaking in each other's natural language.

"Has seh asked for me?" Alyie wondered.

"No. I don't think seh wants to disturb you, but Mom, Thu?glr talks about you constantly."

"I will go there at once." Alyie declared, messaging Hilda to delay her morning appointments.

Riding in a motor car the short distance to the facility that housed the prisoners from B?alam, Alyie rushed into Thu?glr's suite and found shim lying in bed. With the unexpected arrival of the High Governor, the Captain at first thought seh was seeing an apparition.

"It's me, Alyie. I am so sorry I have been remiss in visiting you. My only excuse is the burdens of my office."

"You have a wonderful surrogate in Tessa. She is a delight to be with. How kind she has been to me."

"Is there anything you need from us to make you more comfortable? I don't want you to be in any pain."

"No, Alyie, my time has come. I am thankful that I was able to spend the last part of my life with Humans. You have a lot to teach us if we would only listen."

"Thu?glr, I have to tell you that the High Exalted One Roi Sing?a and I may meet in battle soon. Seh has challenged me to fight shim."

"Be careful, Alyie. Roi Sing?a XV is perhaps the fiercest of all our rulers. Seh is a ruthless warrior who prides shimself on shis prowess. I think it comes from a need to prove shis legitimacy."

"Legitimacy?" Alyie asked. "Is there a question of shis right to rule?

"Oh, Sing?a's bloodline isn't quite pure B?alamian. An ancestor mixed with another species. This has always been a sensitive point with shim. Sing?a overcompensates by being ferocious with all around shim."

“Thank you for the insight. I will keep it in my mind if we meet in combat.”

“I want to give you something,” said Thu?glr, who now was experiencing difficulty in speaking. Seh pulled a necklace off shis neck and presented it to Alyie. At the end of the necklace was a sharp fang.

“This was my first fang. In our tradition, when it falls out of a young one’s mouth, it is kept until a B?alamian makes shis first kill and then presented back to wear forever. I want you to have it.”

“I will wear it with honor and respect in your name,” pronounced Alyie as she placed it over her head.

As Alyie went to gently stroke shis cheek, Thu?glr sat upright and, grabbing Alyie’s right hand, kissed her ring, saying slowly, “You are my Queen.”

“Thu?glr, I am not a queen.”

“You are to me.”

Chapter 25.

In the "evening" before the expedition to Tau Ceti was set to launch, Alyie was busy working in her office. She was in daily conversation on a secure line with Jaynis about the mission. He had decided that instead of an autonomous spaceship to transport the mini-black hole, it would be carried in a human-piloted spacecruiser in which he would be a crew member. Alyie and her Valkyrie wing would take the hyperknot in Teegarden to the Sol System. Once there, they would join a squadron that would make the Beta hyperjump to Tau Ceti. Alyie had sent a video hyperlink message to the respective Regional Governors of Ergo and Cognito of her visit. She assumed these communications would be intercepted and relayed to the B?alamian fleet lying in wait in that system. Once in Tau Ceti, she hoped her presence aboard *Grane* would attract Roi Sing?a's attention. If not, she and her squadron would have to goad shim to attack. After a brief skirmish, she would retreat back towards the hyperknot making sure Sing?a and shis forces were giving chase. While this pursuit is taking place in the Tau Ceti system, the spacecruiser carrying the mini-black hole would be launched and reach the position just outside the Beta knot on the other side of the jump. If the plan works, once Alyie is back into the Sol System, Jaynis would order the black hole to be jettisoned to enter the hyperknot and cause it to gravitationally collapse, trapping the pursuers inside.

There were a lot of conditionals in this stratagem, and Alyie was trying to work through different contingencies, including having to face off against Roi Sing?a. She knew she was placing herself at high risk but believed that her life was worth sacrificing if it could help bring an end to this conflict. Over 600 hundred thousand humans have perished so far during this war, and without some drastic intervention, the death toll would continue to mount. She heard the call of destiny ringing in her mind's ear, and, for the first time in her life, she felt her mortality.

As Alyie was brooding at her desk, she sensed a quiet presence enter the room. Alyie's upper half was stripped down to her tank top with only her burgundy cravat fastened around her neck. She felt exposed, but looking up and seeing it was Reg, she realized there was no need to put her jacket back on.

"Still working on the mission," he said, half-smiling.

"Success comes to those who prepare," was Alyie's attempt at a cheerful retort.

"I was wondering if you are in need of company tonight," Reg asked.

Alyie stood up and came around the desk to stand right before him. Looking her up and down, clad in her white breeches and burgundy boots, Reg gently untied the cravat and noticed the fang hanging on a chain around her neck.

"It is a gift from Captain Kap?stk. He died earlier today. Maybe it will bring me good fortune."

"We could use all the good fortune we can get to pull this off. Are you doing OK?"

"I think I will be alright. I'll decline your offer. I am going back to the mansion to be with my mother and Tessa. I think it is best to spend the evening with them. Jaynis and Tomus are scheduled to do a hyperlink session, and it would be great to be together, at least in spirit, with my whole family."

"As you wish," was Reg's monotone response.

"Don't you want to spend this night with your fellow Panthera? This is what soldiers throughout the ages have done before a battle."

Reg chuckled and said, "Yes, it would be nice to hoist a few and talk about old times."

“Also,” Alyie continued, “I am sure that nice Lieutenant Crosby would want to give you a special send-off.”

“She is Captain Crosby now,” Reg said somewhat timidly.

Alyie raised her left hand to his cheek and planted a soft kiss on the other.

“Have a grand time, Colonel. But don’t oversleep; our ETD is at what serves here as Dawn tomorrow.”

“Yes, Madam High Governor.”

Chapter 26.

The next E-d. the five Valkyric Squadron fighters took off from Teegarden with Alyie aboard *Grane* and Reg seated beside Olga in *Herja*. Making the four E-d. journey to the Sol System via the Alpha hyperknot, they rendezvoused with a small squadron of Federation spacecraft which had broken off from their formation that was enforcing the blockade at L2. Alyie sent an encrypted message to Jaynis, who was now on Mars, indicating she was on course to the Beta hyperknot that would take her to the Tau Ceti system. Upon receipt of the communication, Jaynis alerted his team aboard the space cruiser *Bellerophon* to begin transporting the mini-black hole to the designated position.

Once through the hyperknot, there would be approximately one E-w. trip to reach the location of the B?alamian fleet. The bulk of the enemy forces were on course for Ergo, and the location of the message Alyie received from Roi Sing?a was identified to be within this armada. As her group hyperjumped into the Tau Ceti system, Alyie could not help but reminisce that over 20 E-yr. ago as a Fleet captain, she traveled with Ambassador Felicity Erickson on a mission to Ergo to solve a mystery that involved tachyon messaging into the past. How Alyie wished she could have that tachyon transmitter now and warn her younger self about all the perils that would await the young Alyie. Would her life be any different knowing what the future held? Alyie pondered this as she put *Grane* on autopilot. She resolved that everything that happened to her, good or ill, prepared her for what she was about to face. It was time for the past to meet the present.

Chapter 27.

Musing that Tau Ceti was in the constellation Cetus, a mythological whale, Alyie wondered if this was a foreshadowing of being swallowed by a gigantic beast when her reverie was broken by a message from Olga in *Herja,*

"Intruders coming in fast at 300°; there appears to be a whole wing attacking us."

The alien force must have had their hulls transparent to avoid early detection. Thc battlc had come to Alyie and her escort quicker than they expected. She needed to identify Roi Singʔa's spacecraft and lure shim into chasing her back through the hyperknot.

"Force shields on," Alyie commanded. "Let's take them on a ride."

The Bʔalamian ships started firing projectiles from about 100 km away. The particles ejected were initialed repelled by the protective shields of the Federation forces. However, the overwhelming numerical advantage of the alien ships would eventually rain enough high-energy particles to weaken the shields and denigrate the hulls of Alyie and her comrades' vessels. Alyie's plan was to use the superior speed of her squad's ships to evade the attack long enough until Roi Singʔa came forward to challenge her.

As the "cat and mouse" tactics wore on, there was no sign of the Exalted Ruler amid the attackers. Alyie used the communication link employed when she had last spoken to Singʔa to hail shim.

"Singʔa. This is Valkyrie Starstriker. I know you are out there. Show yourself."

After a short interval, the response came. "Ah, Starstriker. How good to see you in your fine spaceship. Did you think you could make this trip avoiding my fleet? I am surprised you took the chance. You may not live to regret this."

"Is the great Roi Sing?a hiding behind shis great fleet? Here am I the leader of the Humans traveling with a small convoy while you are buried within your mighty armada."

"In due course, I will show myself, Starstriker. For now, we will take our time to weaken your ships. You will soon see the effect of our new weapons. When you surrender, you will see me face-to-face."

Alyie was getting desperate in that her plan of drawing Sing?a out was not succeeding. Seh was correct; unless she ordered a full-scale retreat, the particle beam assault from the B?alamians would force her and her squad to surrender or be destroyed. Playing what she believed was her last card, she declared,

"I can understand why you would not want to be in the front of your formation. As someone who comes from a tainted bloodline, you might doubt the loyalty of those following you."

Immediately after, Alyie heard a curdling roar in her implant, followed by the cry,

"Die Starstriker!"

Breaking away from the alien formation attacking the Federation squadron was a gigantic fighter with the shape of a lion's body and an eagle's head reminiscent of the mythological gryphon, which Alyie presumed to be Sing?a's *K?aal Ch?eeta.* It was soon flanked by a squad of five similar craft. They started firing at Alyie and her Valkyrie teammates even though they were at some distance away. This was the chance Alyie was waiting for. The plan was for the other Valkyrie to provide enough cover for *Grane* to turn and head for the hyperknot, so that when Sing?a pursued, Alyie would have enough of a head start to get through the hyperknot before Jaynis closed it with his mini-black hole. However, the particle beams from these

B?alamian fighters were different than the ones used by the other alien ships. Zarah Khoury, aboard the lead vessel, shouted, "They are using antiparticles; they have eaten through the force shield and are compromising my hull. Retreat, Retreat!"

As soon as Zarah made that emergency declaration, her fighter was imploded due to matter-antimatter annihilation collisions. Bursts of light traveling in opposite directions were seen in its wake. Realizing the dire threat to the mission, Reg broke into the communication system shouting, "Alyie go now. We will hold them off as long as we can."

Alyie was torn between following through with the plan or staying to help her friends. As she wavered, Reg repeated, "Alyie, you need to complete the mission. If not, all we have done is in vain. Please go."

As Alyie watched the three remaining Valkyrie fighters, under heavy assault, fire their laser missiles at the enemy, she saw *K?aal Ch?eeta* turn towards her. For a moment, she considered if she should stand and fight but realized that her defenses were no match for these antimatter particles. Turning *Grane* in a sweeping arc, Alyie charted a course for the hyperknot. As she turned, Sing?a's shot a beam that bore through her shield and struck the drive engine. Still able to power the craft, Alyie noted that the scoop was only drawing in hydrogen fuel at 50% capacity. So much for having an overwhelming advantage in speed over her adversary. Her hope now is that she was far enough ahead of Sing?a that she would still be able to get through the hyperknot before seh did.

Chapter 28.

Deciding that *Grane* needed maximum power to outrun *K?aal Ch?eeta*, Alyie shut off the force shields. She still had a good separation distance on Sing?a, but seh was slowly closing the gap. Systems aboard her ship gave a three E-d. ETA to reach the hyperknot. If *Grane* could maintain this acceleration, Alyie should be able to enter the hyperjump with enough clearance on her enemy's pursuit. She dared not contact Jaynis and Commander Fedor Greysan aboard *Bellerophon* in fear of alerting Sing?a of the trap. Alyie could only hope the ship carrying the mini-black hole was waiting on the other side of the hyperknot when she jumped through. Alyie did try to contact Olga, Reg, and the other Valkyries but received no response. She prayed they were not sacrificial victims to what now looked like her disastrous scheme to end the war.

As Alyie was within two E-hr. of the hyperknot, a jolt was felt from behind. *Grane* was on fire from high energy photons given off by the matter-antimatter collisions produced by a direct hit from a positron beam from *K?aal Ch?eeta*. The radiation emitted seared much of the ship's interior, and only Alyie's protective mesh prevented her from being burnt. Still, she felt the intense heat of the radiation penetrate through her flight suit. Fighting through the pain, Alyie had to leave the pilot's seat and float through the cabin to physically turn on the ship's extinguisher to quench the fire. When she performed a systems check of the spacecraft, it revealed what she had feared. The hydrogen scoop was destroyed; she no longer had any propulsion. *Grane* was now just coasting in space.

"Oh, Starstriker. You thought you could run from me, but I am all-powerful," came Sing?a's heckling voice in Alyie's implant. "You are mine now for the taking. I should offer you the chance to surrender, but your insult was too great. Prepare to meet your God."

Just as Alyie was awaiting the release of the next antiparticle beam, Alyie saw that *K?aal Ch?eeta* was under attack from *Herja,* who in turn was pursued by the other B?alamian fighters in Sing?a's group. In her implant, Alyie heard Reg's voice shouting, "Alyie, we will hold them off. You can make it."

Realizing this was just the diversion she needed, Alyie thought, "Well, it's now literally Newton's First Law or bust."

With reinforcements from shis comrades, Sing?a was soon after able to extricate shimself from the exchange with Olga and Reg's fighter and resume the pursuit of Alyie. However, Alyie had put enough distance between shim that she was able to reach the hyperknot before Sing?a could launch another missile. Going through the jump, Alyie was never happier to experience the sensation of enfolded space and time.

Coming out of the Beta knot, Alyie was relieved to sight *Bellerophon* nearby. She sent a message to Sing?a saying, "I made it through Moggy. Where are you?"

After a brief interval came to the response, "I am coming to get you, infidel."

Receiving a light signal from Alyie, Jaynis gave the order to release the gravity field holding the mini-black hole as his ship approached the mouth of the Beta hyperknot. Once the black hole was ejected, the *Bellerophon* retreated to a safe distance away to join *Grane,* which had maintained its drifting velocity. As the mini-black hole entered the hyperknot, there was an immediate, powerful disturbance in the space surrounding it, and a fierce wave (that Jaynis called a gravitational quake) issued forth. The *Bellerophon* and *Grane,* although by now several hundred thousand km. from the knot, were shaken and turned side-to-side by the ripples through spacetime.

When the wave passed, and the space became stable again, Alyie tried hailing Sing?a, but the signal was dead.

Epilogue

The death of Roi Sing?a did not end the war, but it led the way to its conclusion. Due to the uncertainty of Sing?a's fate inside the hyperknot there was a crisis of leadership among the B?alamians. The Federation sensed this vacuum at the top and seized the initiative. The surviving ships in the squadron, which accompanied Alyie to Tau Ceti were able to land on Ergo. Behind them came a full Federation formation that traveled by a different hyperknot route to engage and defeat the B?alamian Fleet threatening this system. The aliens were also repelled in their attempt to subjugate inhabitants of the Gilese 667 Cc system. Once back in the Sol System, Alyie gave the order for the squadron surrounding the B?alamian forces at L2 to attack, and after a prolonged battle, the Federation prevailed.

Six E-m. after Roi Sing?a's demise, the B?alamians sued for peace. In a ceremony at the site of the G. Soffen Exobiological Institute on Hari in the Trappist-1 system, where the first attack had occurred, a delegation from both species were present to sign a treaty of mutual non-aggression. The event was video broadcast through both Human and B?alamian Space. Representing the Federation were High Governor Alyie Starstriker, recently appointed Vice High Governor Reg Jones, Ambassador and Inter-species Liaison Mateo Santiago, and Queen Revita Jadeite of Kanva. The B?alamian delegation included the new Exalted Ruler Roi Leonidas VII, and shis adjuncts. Conditions of the treaty forbade B?alamian entry into Human Space without special permission by Ambassador Santiago. Any violation would prompt a full-scale invasion of B?alam by the Sidereal Federation.

After the signing, Alyie stepped forward on the hastily constructed dais. She was decked in full formal attire. Unfastening one button of her jacket with her white-gloved hand, she brought forth the fang she wore on the chain around her neck. Displaying it for all to see, Alyie declared,

"This was given to me by Captain Thu?glr Kap?stk, once an enemy, then a captive, but ultimately my friend. I wear this to remember and honor shim. Also, I carry it as a hope for peace and respect between our two species. May this treaty be a first step in reconciling our differences and serving to keep us at peace."

After the ceremony, Alyie and her delegation stayed briefly on Kanva, where they were feted by Revita. Taking leave of her friend, she traveled back to the Sol System, where another ceremony awaited. Reg, Mateo, and Reg's Adjunct Captain Brianna Crosby went on to Teegarden, but Alyie used a shuttle craft on the spacecruiser *Victoria* to land in Denver in what was once the United States of America. Being met by Jaynis, Tessa, Tomus, and Tanita on the landing strip, the family boarded an airjet to take the short flight to Colorado Springs, the site of which at one time was the U. S. Air Force Academy. In the natural canyon, a new multiplex of buildings was nearing completion. Disembarking, Alyie was ushered to the center of the valley in which there was an amphitheater where she would preside over a dedication.

Addressing the dignitaries present, including Alison Bernewithy, Regional Governor of the Sol System, and Silvia Gonzalez, Premier of Earth, Alyie announced,

"By the power given to me as High Governor, I deem this site the new Sidereal Federation Headquarters. The central building which you see to my back is named the Hector Palmero Memorial in honor of my colleague and friend who passionately waited for this day when Earth, the Mother World, could once again take its place as the center of humanity. The High Governor's Office will be on the 10th floor of Hector's building. On Teegarden will remain the Military and Defense Center of the Federation under the direction of Vice High Governor Jones. I am eager to begin the remaining five E-yr. of my tenure as High Governor here on Earth with you and my family."

After applause and congratulations were echoed, Alison came forward to give Alyie a hug and formally cede the site to the Federation.

Soon after, Alyie was surrounded by Jaynis and her other family members, which now included Jaynis' sister, Bernice Bottombrook, and her life partner Saada Nakawa. Tomus, who now sported a pale blue skin coloring, spoke first, "Mom, I understand there is still snow here during part of the year. I will time my visits if Alison, I mean the Madam Regional Governor, allows."

"I definitely want to come when it is winter." Tessa, dressed in her white and brown cadet uniform, declared, "skiing is something I always wanted to do."

"You can only come if you are between sessions at the Academy," Alyie responded, pretending to be stern. "I don't want you taking any unauthorized leaves of absence."

Jaynis and Alyie gazed at each other, she speaking first.

"While we will still be apart with you in Paris and myself in Colorado Springs, at least we are on the same planet now."

"No more need for hyperjumps," said Jaynis.

"No, just time to get re-acquainted," beamed Alyie through green eyes glistening with moisture just before she and her life partner embraced.

----

*From "The Pointy End", Season 1, Episode 8, *Games of Thrones*, Created by David. Benioff and D. B. Weiss, HBO, 2011.

Tessa and Tomus and the Hidden Bough

Deep within all humans lie the beast who can be brought forth when the Master beckons.

José Toro Dura

Prologue

Colonel Arbrit Margiven stood before the 3-D mirror glowing at his reflection. Carlyle Denton, the gentleman's gentleman assigned to Margiven, was finishing dressing the Colonel. A tall, trim man in his mid-50 E-yr., Arbrit Margiven, looked a picture of vim and virility as he stood surrounded by the reflective glass. He had light skin, slicked-down black hair (colored carefully to remove all gray speckles), a sharp angular face with high cheek bones, an aquiline nose, and piercing dark brown eyes. Carlyle had carefully trimmed Margiven's jet-black mustache to give the Colonel a look of style and swagger. He was clad in light brown military-style riding breeches and tall brown synthetic leather boots. The valet fastened a dark brown tie around Margiven's stiffly pressed kaki colored buttoned down shirt. Carlyle then retreated behind the Colonel to help him put on his brown military jacket with stripes and medals indicating his rank and achievements. Then the Colonel's two faux ivory-handled laser pistols were guided into their respective holsters, and the gold belt that secured them was tightened around the Colonel's waist. Carlyle solemnly presented the ceremonial saber that the Colonel had received upon being commissioned. Margiven then inserted it into the synthetic jewel-encrusted scabbard that was affixed to a faux leather belt around his waist. Gazing approvingly at himself, Arbrit donned his cap, put his skin-fitting brown leather gloves on, and struck his riding crop to his right boot to signal to the valet to give a final brush to his boots before dismissing Carlyle.

This was to be a propitious day for the Colonel. He was going riding with Ms. Alison Bernewithy, a woman he greatly desired. It was Arbrit's intention that at the right moment during their time together, he would propose their eternal union. The Colonel was confident that Alison, although she was in a high position in the Sidereal Federation, would not refuse such a personage as himself. He knew her family background, and such a match would only be in her best interests.

From his perspective, having an eternal partner, a woman as important and attractive as Alison, could only serve to enhance Margiven's likelihood of advancement in the Sidereal Federation High Command.

As he strove confidently out of the charming chalet he was staying, following the radial path to the hub building of the Hallows Inn, Margiven was delighted to see that the sky on Cognito was clear and the air was fresh and clean. It would be a perfect day for a ride. The planet, a member of the Gilese 667 C system, had been terraformed some E-c. ago and now had a temperate climate with leafy forests, green valleys, lakes, rivers, and breathable air. (The planet had an orbital period around its red dwarf star of about 90 E-d., and a rapid rotation that produced day and night on a 20 E-hr. cycle.)

The Colonel had convinced Alison they should take some time off to get to know one another better. Alison was Regional Governor of the Sol System stationed on Earth in Brussels and the Federation's Director of Arts and Culture while he was the Superintendent of Federation Fleet Academy in New Cedonia in the Rebus Centauri system. They had first made contact through an AI social matching algorithm that guaranteed their mutual compatibility and had met in person when he had traveled to Earth on Fleet Command business. Alison had given him a tour of Paris, and they went out several times during his one E-wk. stay. From their conversations, he and Alison discovered they both shared a common interest in horseback riding. Margiven chose Cognito as a perfect place for them to get to know each other better and also partake in the planet's amenities. Far enough away from the center of Federation activities, Cognito had developed into a destination place for those who wanted to indulge themselves away from unwanted scrutiny. The Hallows Inn where they were staying promised all forms of luxury, including for each guest a personal human staff member to attend to all their needs. If this day would go as the Colonel planned, he would not be spending the second night in Cognito alone in his room.

Soon after Margiven was seated for breakfast, Regional Governor Bernewithy arrived bedecked in her riding attire. Her chalet was on the other side of the Inn, adjacent to the artificial Lake Cithaeron. Alison was a petite woman in her late 30s E-yr., with white skin, yellow hair pulled back in a bun, and radiant blue eyes. Her face was powdered to look pink-cheeked, and she had applied a generous amount of bright red lipstick. She was clad in a dark blue riding coat, and underneath was a starched white button-down shirt fastened at the collar with a flowing white bow. Alison had also secured a sky-blue scarf around her neck. Her white breeches were tucked snuggly into tall, highly polished black boots.

The Colonel rose, clapping his boots together, as Alison approached. "Good morning Mademoiselle Regional Governor. Je suis enchanté."

"Good morning Colonel Margiven, and, by the way, you look pretty dashing yourself in your full military regalia."

"I hope we can dispense with the formality; please call me Arbrit. May I call you Alison?"

"As you wish. But first things first, I am hungry; I hope you are also."

The Colonel tapped his implant to signal the autobots to serve the breakfast, consisting of wheat cakes, synthetic eggs, cultured meats, fruit, cream, and assorted drinks such as juice, coffee, and tea.

After the hearty meal, both Arbrit and Alison excused themselves to their respective lavatories. Once done, they strolled outside down the pathway to the stable, where their mounts were already saddled and held by their respective grooms. The horses were selected by the Inn's algorithms based on a client's desired gender, color, size, and temperament. Alison's mount, Harmonia, was a docile milk-white mare of average height, while Arbrit rode Zagreus, a tall spirited chestnut gelding. Inside the enclosure, they donned protective helmets and put spurs on their boots. Feeling his nano-molecular enhancers coursing through his bloodstream fully activated, Arbrit grabbed Alison just as she was about to mount her horse. He held her tightly against his body as he kissed her fully on the mouth. She reciprocated by returning his kiss, albeit with less ardor. Feeling sure this would be a good day, the Colonel reluctantly pulled himself away from Alison and helped her mount. After getting onto his steed, Margiven herded Alison out of the barn. Leaving the stabling enclosure, the couple galloped together into the adjoining meadow.

After an invigorating ride for about two E-hr., they returned to the paddocks just as it was starting to get dark in the short Cogitian day (the planet's axis being pointed away from the star at this time of its cycle). After helping Alison dismount, Arbrit could hardly restraint himself as he embraced her passionately.

Gently pulling herself away, she decried, "Colonel, in my position, I need to keep some sense of propriety, as so should you. Let's take this inside. I will tell my personal attendant, Nina, to bring a bottle of champagne to my suite."

After leaving the stable, Arbrit Margiven and Alison Bernewithy walked around the circumference of the Inn's grounds to Alison's elegantly appointed chalet. Entering, Alison excused herself as Arbrit waited for the champagne to be promptly delivered by Nina Mori, a slim but tall attractive young dark haired woman. When Alison emerged from the lavatory, she had taken off her jacket, untied her bow, placed her scarf on the side table, took off her gloves, and gotten out of her boots.

Seeing the Colonel still fully clad in his uniform, she laughed, "Do you always stay dressed in the boudoir of a lady?"

"A toast," proclaimed Margiven as he poured the effervescent liquid into the two flutes provided, "to our union."

"Union?" said a quizzical Alison as she sipped her champagne.

"To being here together then," was Arbrit's rebut to her question.

"I can drink to that," said Alison as she used her implant to discreetly turn down the room lightning and close the shades to inhibit daylight.

As soon as she emptied her flute, Arbrit could hold back no longer, almost violently removing his holster belt, scabbard, and jacket. He came to Alison, gloves still in his hands, kissing her wildly about her face. Alison responded readily, and soon they were entwined together. After unfastening each other's breeches and belts, the end came quickly as Arbrit felt the pleasure of release while Alison held him tightly.

After their comingling, the Colonel refastened his breeches and re-tightened his belt while Alison tried to tidy herself up. Suddenly a sound was heard near the full-length window overlooking the patio. A black hooded and robed figure had apparently snuck into the room, bypassing the security at the patio entrance. After Alison had re-activated the lights, the intruder could be seen wearing a stone-like mask consisting of holes for eyes, nose, and mouth and fake curls and grapes surrounding his/her head and beard. In the intruder's hand was a staff entwined by vines and leaves. There was also a distinct smell of licorice now in the room. With eyes darting to and fro, the stranger spied the scarf and, with a single motion, whisked it off the table, twirling it around the staff. Retreating his/her steps with the grace of a dancer, the intruder left the room whence she/he came via the patio.

"Well, that was an unexpected climax to the day. Do you think this is all part of the Inn's entertainment?" said a somewhat startled Alison. "My only regret is that scarf was pure silk. I doubt I can afford to replace it."

Arbrit's face had turned bright crimson red. "The outrage! To invade our privacy is abominable. This cannot stand," shouted the

Colonel as he hastily put back on his jacket and fastened around his waist the laser holster and scabbard.

"Calm yourself, Arbrit. I am contacting security. They should be here soon to take our statements."

The Colonel would not be pacified as he went outside to the patio. Standing at the outskirts of the glade that overlooked the lake was the intruder waving his staff with the scarf, seemingly taunting the Colonel.

"Why that blackguard? How dare he insult me!"

The Colonel then drew his laser pistols out of their respective holsters and was preparing to fire.

"Don't shoot, Colonel. This is not worth killing for."

"He needs a good thrashing," said the Colonel as he returned the pistols to their holsters and drew his sword. "I'm going to teach him a lesson he'll not forget."

Margiven then bounded out the patio towards the stranger, who, when the Colonel came close enough, withdrew into the glade. Margiven quickly followed, brandishing his sword high overhead.

Alison was finishing putting her boots back on when the security team arrived in her chalet. Together they went through the patio towards where Alison had seen the Colonel enter the glade.

As they approached, cries were heard, and light flashes were seen in the forest. Just as Alison and the staff members were near the glade, the Colonel staggered out, buckling under his boots. From the light beams, the security officers shined at him, it was clear Colonel Margiven was gravely wounded and bleeding profusely, staining his jacket and breeches. He appeared to have been hacked at and maimed in the chest, arms, and legs, and there was a gaping wound in his neck which appeared to partially sever his head. Alison rushed to Arbrit's side, unbuttoning his jacket and unknotting his tie to make the Colonel comfortable as the security team rushed to get medical aid. Holding his right hand tightly, Alison cried softly as she saw the life slowly ebb from her gallant Colonel. He died in her arms with his boots outstretched on the grass.

One E-yr. later

Chapter 1.

It was a busy day in the archives of the Louvre. Tomus Starbrook, Assistant Archivist, Part-Time, was cataloging art objects for an exhibit on arcane symbols of ancient man. The Museum's Curator of Ancient Artifacts, Dr. Nasu Godahfi, had tasked Tomus with this assignment. It was tedious work scanning through the Louvre's vast collection spanning the breadth of human history, but Tomus was fascinated with the treasure trove he was cataloging.

Now 18 E-yr., the son of High Governor Alyie Starstriker and Dr. Jaynis Bottombrook, Chief Scientist for the Sidereal Federation, was a fair-skinned, lanky young man with spiked dyed-orange hair and bright blue eyes. He was dressed in his usual loose-fitting pullover sweater, undershirt, synthetic cotton pants, and soft elastic shoes. Tomus was technically a student at the Université de Paris, but he gave all his spare time to his work at the Louvre. It was convenient in that he was staying in a pied-à-terre in Paris that originally belonged to his father and Alison Bernewithy. With Dr. Bottombrook's move to join his spouse at the Federation's Earth Headquarters in Colorado Springs, Alison had allowed Tomus to share the apartment with her. She had recently stepped down from her position as Minister of Arts and Culture to devote full time to her duties as the Regional Governor of the Sol System. Although Alison usually resided at the new Regional Governor's suite in Brussels, she remained in touch with Tomus. When she was in Paris, they often dined together and shared news about what was going on in the art world.

What struck Tomus peculiarly this day was that when combing through the archives he found images of the mask representing the ancient Greek god Dionysus. It was the figure of a man with curls and a beard made of vines and grapes – identifying him as the god of wine. Tomus immediately recalled Alison's description of the intruder who was the instigator of the murder of Colonel Arbrit Margiven on Cognito one E-yr. ago. The case was handled by the local authorities who claimed planetary jurisdiction under the Act of Self-Determination because although Alison and Colonel Margiven were Federation officials, their trip to Cognito was purely personal. (The fact they were having a tryst when the break-in occurred made the Federation reluctant to publicize the attack.) This resulted in a less than thorough investigation in which no suspects were identified, and of course, no arrests were made. The intrusion was ruled an attempted robbery in that the Colonel, although mortally wounded, was able, by his actions, to foil the thief. (The stolen silken scarf was recovered after a search of the glade where Margiven was assaulted.) The Colonel was accorded a full memorial service on New Cedonia, befitting a hero of the Federation.

From Tomus' conversations with Alison, he knew she was not satisfied with this conclusion, arguing that the assailant's intent was the assassination of the Colonel and herself. Just to be safe, Tomus' mother, Alyie Starstriker, had commissioned two of her Elite Guards to provide protection for Alison. Now saddled with having them in her proximity constantly, Alison realized the luxury of private life had ceased, and she naturally grew suspicious of new acquaintances.

Chapter 2.

Knowing how upset Alison was by the premature conclusion reached on Cognito of this horrific event, Tomus decided to pursue it further. His first step was to contact a person he knew who also had an interest in the case – his twin sister Tessa. She was in her third year of the Fleet Academy on New Cedonia, and she had personal knowledge of Colonel Margiven, given he had been the Academy's Superintendent.

Gauging that it would be technically evening on New Cedonia (an artificially constructed base orbiting a moon in the Rebus Centauri system), Tomus hyperlinked his sister and found her in the rec room carousing with some other cadets.

"Hi, Tomus; what brings this call from my hard-to-reach Earth-dwelling sibling? Want to visit me in New Cedonia? I will take you for a ride on the latest fighter. It can go from 0 to 5 g in 20 E-sec. It's, as they say, a blast."

That last comment produced raucous laughter and hoots from Tessa's apparently intoxicated comrades as they seconded her invitation. Tessa, herself, was holding what looked like an old-style flagon of some liquid that was spilling. As she swayed to and fro to syncopated background music, she looked at the picture of a young Earth demi goddess clad in her Academy cadet tight-fitting uniform of white turtle neck pullover, snow white pants, and brown leather-like boots. Tall and athletic with bright green eyes, pink cheeks, and freckles, she was the image of a younger version of her mother, except Tessa's hair was more golden than red.

"I need to talk privately with you on a serious matter."

"Oh, Tomus, everything with you is serious. These are my best buddies. Anything you want to say to me can be said in front of them. Unless it is of a 's', 'e', 'x', 'u', 'a', 'l' nature."

"It involves the murder of Colonel Margiven."

The late Superintendent's name got the attention of Tessa and her friends.

"Poor old Arbrit," said a grinning Argus Fitzhugh, sticking his red-haired head so he could be seen in the video screen. "He was always so pompous. It was time he received his just reward."

"At least he was able to get one last shot before the axe fell. He died in his boots, but not until after he had to put his pants back on, I understand," offered a laughing Naja Kwan, a young woman with black hair and pale white skin.

"Tessa don't you and your brother know the person he was with?" asked Objie Narubi, a tall cocoa-hued woman with a curly-haired brown mane. "What's her name?"

"Alison Bernewithy," volunteered Tessa.

"Isn't she the Regional Governor where you are stationed, Tomus?" asked Argus. "Tell me, is getting a rise being with her worth getting killed over?"

Tomus became flush from that last remark, and he was about to blurt out some incendiary response, but Tessa seeing how upset her brother was getting, interjected, "You were right, Tomus. Let's continue this conversation later. I will contact you when I return to my bunk."

Chapter 3.

When Tessa was alone in her room, she used her implant to reach Tomus. He was still in the subterranean level of the Louvre, where the archives were historically stored. Tessa could see from his image her brother had calmed down, but he was still upset over her friends' crude humor.

"Why do you associate yourself with such louts? Mom and Dad would be appalled at the company you keep."

"Oh, Tomus, they are not that bad. A little too much drink can produce some forgettable behavior. You caught us when we were having some downtime. You should try to have some yourself once in a while."

"A man who was the head of your Academy was horribly murdered, and a fine woman who has cared for us since we were toddlers, we have been terribly traumatized by this crime. You and your friends should show some respect and sensitivity."

"My mates are not malicious when they are sober. Besides, these are the people I will depend on when we go into combat. I trust them with my life. Still, Tomus, you are right. Colonel Margiven, no matter what his personality quirks, was a distinguished soldier and an able Superintendent. His loss needs to be mourned, not ridiculed. And yes, Alison has been involved in our family for many E-yr. I know she saved your life once and how much she means to you."

"The reason I have contacted you is that I have discovered something that may be related about the case from looking at photos of arcane art."

"Brother, I think you have been in the basement of that museum too long. How could ancient artwork give any clues to the murder of Colonel Margiven? Maybe there are some fumes where you work that have affected your brain."

“Tessa, my mind is as clear as ever. Listen, there have always been inconsistencies in this case. The Cogitian authorities concluded it was a robbery, but all that was taken was a silk scarf, which, by the way, was recovered.”

“Dear Brother, do you know how expensive real silk costs on the Common Exchange? Ask your friend Alison. I am sure she knows the value of that scarf.”

“Also, Tessa, the intruder, after snatching the item, didn’t just escape into the woods but stood outside displaying the scarf and daring Alison and the Colonel to give chase. He or she was trying to lure them into the glade to attack them. This was not an attempted robbery but an assassination.”

“I am still not convinced. Sure, the stranger was taunting the Colonel. But when Margiven came out flashing his sword, vowing vengeance, the thief in the form of self-defense, managed to overpower the Colonel and mortally wound him.”

“That doesn’t make sense. The wounds Colonel Margiven incurred were not from a saber but from axes and spears. Besides, do you think a skilled soldier such as the Colonel could be bested by a single individual? The attack was made by several accomplices of the thief in the forest lying in wait for Margiven, and perhaps Alison, to pursue. The way she described the Colonel’s neck wound was that whoever did it was trying to decapitate the Colonel.”

“Are you saying this was not a simple murder but a premeditated plot?”

“Yes, and what I have found in the archive collection helps to support this hypothesis.”

Displaying the image of the mask he had discovered for Tessa to see, Tomus explained, “This is a representation of the ancient Greek god Dionysius, and it is very similar to Alison’s description of the mask worn by the intruder.”

“So the intruder put on a mask of some archaic Earth deity. It was probably just a convenient way to disguise him or herself.”

“But there are some other concurrences that associate this god with this crime. Do you know anything about the cult of Dionysius?”

Is this a test, Tomus? Let’s see. If I am not mistaken, Dionysius, or Bacchus in Latin, was the god of wine and fertility. He is most associated with wine festivals, dancing, and frivolity. Seems like a fellow who would fit right in with my crew.”

“But there is a darker side to the worship of Dionysius. His cult began in about 1500 BCE. His followers were known for achieving religious-like ecstasy through intoxication in a state called *bakkheia* that resulted in such a frenzy that they would commit violent acts. They believed in this hallucinatory condition that they were Dionysius, and through his power, they were known to tear animals and even people apart.”

“So you are suggesting Colonel Margiven was a victim of this cult.”

“Yes, I believe he and Alison were deemed sacrificial victims in a reenactment of an ancient blood ritual.”

“Wow! This is way out there. We are living in the 26th (CE) E-c. How can the rituals of worship some 4000 or so E-yr. ago have any connection to a simple murder over a stolen scarf?”

“I think there is enough evidence to open a new investigation.”

“But the Federation is not allowed due to the Act of Self-Determination.”

“You are correct, Tessa; the Federation is constrained, but you and I are not.”

“You want us to solve Colonel Margiven’s murder?”

"Yes, he was your Superintendent, and Alison is almost a part of our family. We owe it to both of them."

"I do not know Tomus if this is worth our time and the repercussions we would suffer from the rules we would have to break to accomplish this."

"Ok, let's start with small steps. I want to question Alison in detail about what she remembers. She is coming to Paris next E-w. to visit the Museum. I will arrange a meeting with her. Meanwhile, can you use your influence to see the case files of the murder on Cognito?"

"Influence? What influence do I have? I am just a cadet on New Cedonia."

"Yes, but Tessa, you did work for the Department of Federation Security at one time. There must be someone you know who will give you access. You are the High Governor's daughter, after all."

"I have a funny feeling this is going to get us into trouble. What will happen if Mother finds out?"

"I trust you can be discreet enough that she need never know. After we complete our assignments, we can share what we have found and then decide if there is enough evidence to pursue the case further. Agreed?"

"Brother, this is a side of you I have not seen before. I think it becomes you. You're on, partner."

Chapter 4.

Next E-d. Tessa hyperlinked Finneas O'Rourke, a young man she interned with before she gained acceptance into the Fleet Academy. O'Rourke was always friendly, and Tessa suspected he had a crush on her.

Making sure her clothes fit just right and her golden locks were lustrous, Tessa beamed, "Hi, Finneas. It's been a while. How are you doing?"

"Why, Tessa Starbrook, hello back. You look amazing. How are things in flight school?"

"Just fine, Finneas. I have been thinking of you of late and some of the fun we had together. How is you doing at the Department?"

"You have? I am now Officer, Third Grade. When Vice High Governor Jones did a recent review of the Department, I was told he saw potential in me."

"That's great. I am planning to come to Teegarden soon, flying the Federation's new Xi class fighter. When I come, would you like to take a ride with me?"

"Sure, Tessa. I didn't know they let cadets pilot the newest spacecraft."

"It helps to know people in high places, Finneas. This brings me to ask a small request of you."

"Anything for you, Tessa."

"I have been tasked with a special mission to look at certain crimes across Human Space. I need top clearance access to the Federation's Security System. Can you provide that for me?"

"Tessa, I don't know. Why does a cadet need to look at reported crimes? Where are you planning to look?"

"I am sorry, Finneas, but I am not allowed to divulge details."

"Tessa, I really can't give you access. It is forbidden to all except those on the staff of the Department."

Tessa forced a frown, pouted her lips, and knitted her brows. "Are you saying you can't or won't give me clearance?"

"I am so sorry, Tessa, but I am not authorized."

"Ok, Finneas. Then I will take my request to Vice High Governor Jones. I didn't want to disturb Reg because he often tells me how busy he is. I am sure he will take note of your refusal to grant me this clearance."

"Please don't do that, Tessa. I will give it to you."

"Thank you, Finneas. Your assistance is appreciated. Goodbye."

Chapter 5.

Tomus was seated at a corner table in Le Café Richelieu Angelina in the Louvre near a large window looking out to the five E-c. old pyramid that sits in the Cour Napoléon. Café Angelina is a time-honored place for fine cuisine food (especially French pasterries) and conversations among Parisians. He was waiting to have dinner with Alison. She had been all day at the Museum in meetings with officials in the Ministry of Arts and Culture. Even though Alison had stepped down from her post of Director after she was elected for a full term as Regional Governor, she was still very involved in the Arts.

Self-conscious of his casual appearance, Tomus was admonishing himself for not dressing better for this tête-à-tête when he spied Alison down the corridor coming towards him, making loud noises as her very tall black high heels clumped on the hard stone floor. She was dressed in a smart business suit, black jacket, and skirt. Her Regional Governor insignia was pinned to the left lapel of her jacket, and underneath, she wore a cream-colored satin blouse fastened at the collar by a white bow. Her mid-length blonde hair was held back by a large amber headband. She had put on some weight since Tomus had last seen her in person, but to him she looked radiant. (Trailing behind Alison were her ever-faithful guards trying but not succeeding in being unassuming.)

As Alison approached, Tomus rose up to greet her. Breathless, she hugged him warmly, kissing him on the cheeks and leaving bright red lipstick marks.

"Tommy, it has been too long, and we live so close to each other. This was a great idea to have dinner together. I wish I could stay overnight, but I have an early meeting at HQ tomorrow morning. Please, let's sit."

When the waiter came, they were served a dinner of poisson, asperges, pommes de terre rouges et salades fraîches. Alison ordered a bottle of Chenin Blanc for the table.

"I know this wine pales compared to the ones your father produced at Montpellier, but I think it will suffice for us. À votre santé!"

Enjoying the meal and especially the chocolate chaud afterward (which Alison also ordered served to her guards), Tomus and Alison shared stories of their respective work; he about his studies in human art history, and she about the new Regional Headquarters complex in Brussels. Alison was also very excited about the new initiatives planned for expanding the Arts and Culture Ministry on Earth.

"I know I am not Director anymore, but I can't help getting involved in all the great things happening on Earth involving preserving human history and culture. You could say that is where my heart is."

"Yes, I have read that new Art Centers are opening in Shanghai, Buenos Ares, and Melbourne," volunteered Tomus. "Earth now has so many visitors."

"This is no doubt the result of your mother moving the Center of the Sidereal Federation to Earth. Her presence here also is a boon. People want to just come to the planet she walks on, in the oft chance they may see her."

"Mom's term as High Governor is coming to an end in two E-yr. Are you thinking of running for it?"

"Tommy, I pity anyone who will be elected to replace your mother. Whomever it is will never be able to live up to the standards she has set. No, I am content for now to be the Regional Governor of Sol. Earth is in my blood. But enough about me. Tell me about your research, Tommy."

"Before I do that, I want to question you about the events on Cognito. I know this is painful, but I may have some new information that can help you understand what happened."

"Painful is correct. I keep trying to replay everything that happened that night. I should never have gone to Cognito. I don't know what I was thinking. Colonel Margiven was not for me. Yes, he was dashing and dynamic, but I knew it was only going to be transient. It seems I am doomed to failure in matters of the heart."

"I don't feel that way. I know I am young, but people say I am wise beyond my years. If you have patience until I have finished my studies, we can live together in Paris."

"Oh, Tommy, you are so sweet and undoubtedly the finest man I will ever know, but we can't be together. The age difference is real; you are just 18, and I am going to be 40. I value your parents' friendship too much to risk losing it. I have already been involved enough in the affairs of your family; I think my quota is used up. What is your new information?"

"I think that the death of the Colonel was a ritualistic sacrifice by an ancient human cult. You were lucky you were not part of it."

Tomus then proceeded to tell Alison about the mask of Dionysius he found in the archives. When he showed her the picture, Alison confirmed it was just like the one the stranger had on who invaded the room.

"Was there anything else you remember about the intruder's appearance or behavior?" Tomus asked.

"There was one other funny thing that I didn't tell the hotel security because it seemed frivolous. There was a scent of licorice in the room soon after the stranger entered, and it went away after the intruder left."

“Alison, this is important. You have told me that the intruder was carrying a stick or staff that had leaves and ivy wrapped around it.”

“Yes, he or she used the staff to take my scarf off the table.”

“Dionysius or one of his acolytes were known to carry a scepter made of fennel, called a thyrsus. That thyrsus was the source of the licorice smell. This can’t be a coincidence,” exclaimed a triumphant-sounding Tomus.

“Tommy, what are you going to do with this knowledge?”

“I am going to find out who threatened you and killed Colonel Margiven.”

“Be careful, Tommy. I will be feeling even more wretched than I do now if anything should happen to you, dear boy.”

“Don’t worry, Alison, I have an able partner to assist me.”

“And who would that be?”

“My sister.”

Chapter 6.

At a pre-designated date and place, the twins hyperlinked to share the information each had gathered. Tomus was eager to tell Tessa about the presence of the thyrsus, a sure sign the intruder was masquerading as Dionysius. Tessa, in turn, had details about the investigation of Colonel Margiven's murder on Cognito.

"When I was able to access the file, which had been sealed, I found there was evidence in the glade of a struggle, and from the way, the brush was disturbed and the number of boot and shoe prints in the ground, Margiven was attacked by several assailants. So, you were correct, Brother dear, in claiming the theft of the scarf was just a pretext to lure him to his death. The Colonel didn't go down without a struggle. There was blood on his saber, and one of his laser pistols had been discharged. The images of his body were gruesome. There were at least 15 entry wounds, and if he had not managed to break free and stagger back into the clearing, they would have finished decapitating him."

"Is there any evidence that Alison was also targeted?" Asked Tomus nervously.

"Not so much, other than the fact that if she had joined the Colonel in pursuit, I think we can presume she would have been attacked as well."

"Why were the authorities on Cognito so quick to conclude it was a robbery attempt?"

"They wanted as little publicity as possible and certainly did not want the Federation sending some investigative task force to Cognito, disrupting the Inn's tourist business. Alison gave them eyewitness testimony of the theft of the scarf and could verify that Margiven was intent on retrieving it. Other than the costume worn by the stranger, there were no other leads. So based on this scanty evidence, the authorities made their conclusion."

"My feeling, Tessa, is that there is a lot more to be found at the site of the crime. What do you think?"

"I am inclined to agree with you, Tomus, but this crazy rite may not be just on Cognito. After reading the report, I used my clearance to search for similar crimes investigated in the last few Earth-years in other places in the Federation. My search found hits in Ergo, Kanva, and Nespar. In all cases, there was a dismemberment: a person on Ergo, but in the two other locations, the sacrifice was an animal. All the deaths occurred in leafy woodlands."

"That indicates this cult is widespread. But I really think we need to get over to Cognito and see for ourselves what happened that night."

"How are you and I going to be able to travel to Cognito? We don't have any means or authority?"

"This will take some planning and a certain amount of deception. Tessa, I will figure it out."

Chapter 7.

An E-w later, Tessa was having dinner with her grandmother, Tanita Starbright, in the latter's apartment on New Cedonia. When Tessa began her candidacy at the Flight Academy, Tanita assumed a senior faculty position. This was arranged by her daughter, Alyie Starstriker, who told Tanita she wanted her to be near Tessa to provide support. No doubt, the real reason was Alyie didn't want her mother to be alone since her husband's (and Alyie's father's) death five E-yr. ago. Tanita, now a septuagenarian, had a fine chiseled face and hair which had gone a brilliant white and dark brown eyes. This evening she was dressed in a dark purple pants suit with side buttons on her jacket extending high up her collar. She glided around the apartment wearing felt black loafers.

After they finished eating, Tessa and Tanita had a hyperlink video session with Alyie, who was in her office at Federation Headquarters in Colorado Springs.

"Hello," chirped a smiling Alyie, who was dressed in her characteristic white jacket with the High Governor insignia worn over a bright red pullover blouse. Her face showed the effects of the responsibilities she bore, but, all in all, the High Governor was still a spectacular-look woman. "Ladies, how are you doing at New Cedonia? I have such fond memories of my time both as a cadet and the years I was stationed there with my Valkyrie Squadron."

"Well, Mom, the food still sucks, and the instructors are very demanding. The rules are so strict that just to visit Grandmother, I had to get special permission and state this would be a counseling session."

"I am sorry about the food, but on Fleet missions, trust me, it will be much worse. But I am pleased to hear about the rigor of the faculty. I am also grateful that you have not been given any special treatment."

"But Alyie," interjected Tanita, "I should be able to see my granddaughter without stretching the truth to do so. Don't you think that is reasonable?"

"To be honest, Mother, the more Tessa conducts her service for the Federation without special consideration, I think it will be better in the long run for her, but I will allow you to request her presence whenever you feel the need to see her."

"Thank you, Madam High Governor," said Tanita affecting mild sarcasm.

For the rest of the session, the three women shared stories of food, friends, and family.

With the hyperlink call concluded, Tessa approached Tanita with her primary reason for the visit.

"Grandmother, Mom's 50 E-yr. birthday is coming up in less than a year, and Tomus and I are planning a surprise."

"Great! Can you tell me what it will be?"

"Sure. And you can help make it happen. We would like to present Mom and Dad with a gift of a trip to an exotic place where they can be together, free from the hassles of their Federation duties."

"That sounds wonderful, Tessa. Although Alyie and Jaynis are both in Colorado Springs, I doubt their respective duties allow them much personal time together. Where are you and Tomus thinking of sending them?"

"We are not sure yet, but we want to visit some possible sites in person to see what they have to offer."

"How can you do that with Tomus in school on Earth and you here at the Academy?"

"That's where you can help Grandmother. If you assign me a special project that allows travel, I can take Tomus with me. His studies at the Université are flexible, and I am sure he can get time off from the Museum."

"I do not know, Tessa. I would have to get permission from the Academy Board. If your mother finds out, I am sure she would not approve of this "special treatment".

"But Grandmother, it's for her and Dad. Since Tomus and I want to make it a surprise, she won't know beforehand. We need to keep it secret. You are on the Board; I am sure you can make the case to let me go."

"When will this happen?"

"As soon as you give me approval. A plus factor is you will get to see Tomus since he will fly to New Cedonia before we depart together."

"It will be wonderful to see Tomus. I miss him so much. Do you know where you will travel to first?"

"We were thinking of Cognito. We have heard that there is a beautiful inn on the planet that provides privacy and luxury to its guests."

"Isn't that where poor Colonel Margiven met his end trying to retrieve some stolen item? I understand it was gruesome."

"Oh, that was long ago. I am sure there are now protections against that sort of violence. Thanks so much, Grandmother, and remember no one but myself, you and Tomus need know."

Chapter 8.

As Tomus was getting ready to leave the Museum prior to his departure to New Cedonia, he stopped off at the office of Dr. Nasu Godahfi to pay his respects and thank the Curator for granting him leave to visit his sister and grandmother. Dr. Godahfi was sitting in his chair behind his large desk, inspecting a parchment with an old-fashioned magnifying refractor. The Curator was a usual, fastidious in his appearance and attire. The 60 E-yr. old gentleman had grayish hair carefully coiffured, olive skin tones, and a well-manicured gray beard. He was dressed in a tailored dark blue and white pin-striped suit, and he had knotted around his starched white shirt a light blue silk-looking tie.

"Have a good trip, Mr. Starbrook but don't forget to keep up your studies. The more you learn, the more valuable you will be to us here in the Museum."

"Yes, Dr. Godahfi. I have uploaded all the pertinent materials to continue my curriculum. I was wondering if you have time to answer some questions that have come from my archival work."

"Of course, Mr. Starbrook, ask away."

"The ancient gods of Earth's Greek and Roman culture faded away with the advent of Christianity. How did that happen?"

"It is much more complicated than that, and if I spent the time to explain it all, you would miss your flight. Here is the short version. Before Christianity, there was Judaism, the first prominent monotheistic religion that unified belief in a single deity which seems more rational than a zoological collection of supernatural beings. But you are correct in claiming the Christian religion in the Roman Empire supplanted belief in the many gods."

"Were there any societies that clung to these ancient gods, resisting Jewish and Christian teachings?"

"You should also include Islam which in many ways was more effective in erasing previous belief systems. Please remember we are talking about many Earth centuries for this evolution to take place. Even in our era it has taken a long time for organized religion to lose its grip on society and become more personalized."

"But Dr. Godahfi, are you aware of any groups throughout the Federation currently practicing worship of the ancient gods."

"The old gods, as they are sometimes called, demanded things we abhor today, such as blind faith, monetary offerings, and even in some cases, blood sacrifice. I am sure there are some individuals who despise the Federation for its reliance on realism, order, and scientific understanding of phenomena. They are attracted to the mystery and awe the worship of these gods engenders. They put all their trust in these imaginary higher beings, which they call faith. These acolytes believe humanity was not meant to explore space, colonize planets, and use scientific knowledge to alter the Universe for its own benefit. Rather than deny our bestial heritage, they feel we should embrace it."

"But you are not aware of any cult active now?"

"I have heard rumors, Mr. Starbrook of a group who, using a discredited 20th (CE) E-c. philosophy called Anthroposophy believes they can channel the spirits of beings from Earth's ancient past. But that would be so retrograde and archaic in the society we live in now. Most psychologists believe such worship is just a manifestation of the worst primal instincts in humankind. We have ascended from vague fears about our existence on Earth to strong confidence concerning humanity's role in the entire Universe. No, there is no place today for such a group."

"Thank you, Dr. Godahfi."

"Have a good visit with your family, Mr. Starbrook."

Chapter 9.

After Tomus made the trek to New Cedonia, he spent his time on the base studying while mostly staying with his grandmother. Tessa, once securing approval for the excursion to Cognito, made arrangements to command a spacecruiser for the journey. On the day before their departure, Tessa received a message on her implant that Captain Brianna Crosby, adjunct to Vice High Governor Jones, would like to speak to her. Tessa had been an assistant to Captain Crosby when they both worked at the Department of Federation Security. Since Reg Jones had become Vice High Governor, the Captain transferred to serve under him. It was widely known she and the Vice High Governor had more than a professional relationship.

"Tessa, hi. How are you doing at the Academy?" announced a smiling Brianna once Tessa had returned the hyperlink call. The Captain was a young woman in her late-20s with a beautiful caramel complexion, full figure, tight-knotted black hair, and almond-shaped yellowish eyes. The video image showed her dressed in a figure-flattering pale yellow blouse, fastened at the collar by a black ribbon tied in a neat bow.

"I am doing great, Brianna. It's wonderful to see you. What can I do for you?"

"Reg, I mean the Vice High Governor, asked me to contact you. Hold on, I will send your call to him."

Tessa waited nervously for the transfer to be completed. She suspected Reg had knowledge of her and her brother's scheme and would thwart their plans. Tessa needed to think quickly to convince Reg of the need to go to Cognito.

"Hello, Tessa," came the Vice High Governor's bass voice. He had aged a little since last she saw him, but he was still the picture of a man of power and self-confidence.

"Hello, Reg," answered Tessa guardedly.

"I understand you and Tomus are making a special trip to Cognito. Why, of all places, do you need to go there?"

"It is to be a surprise, but if you promise not to tell anyone, Reg. I'll let you in on it."

"Tessa, you are on thin grounds getting a commitment from me, but I will indulge you a little."

"Tomus and I are scoping out sites for a surprise 50th birthday trip for Mom."

"But why Cognito?"

"Oh, Reg, isn't it supposed to be an ideal location to escape to?"

"It is also the scene of a gristly murder of a Federation Colonel that also involves a Regional Governor that both you and Tomus know very well."

"Oh, but it might be perfect for Mom. I feel we need to go."

"I am going to allow it. But you will be accompanied by two Panthera who are stationed on New Cedonia. Also, I have assigned a very experienced commander as your co-pilot."

"Why do I need to have a co-pilot and some guards? We are just looking for a venue."

"Listen, you can give me that innocent smile in that pretty face so much like your mother's, but I am not such a fool as to not see through your ruse. My conditions are non-negotiable. Either you have the escort, or you stay on New Cedonia. Make your choice."

"Well, Reg, if you put it that way, they are free to join our junket to Cognito."

“Make sure you stay safe, if not for your sake, than mine. If anything happens to you or Tomus, imagine how I would feel explaining this to your mother.”

“Will do. Thanks so much. Give my love to Brianna.”

Chapter 10.

As Tessa and Tomus arrived at the spaceport with two autobots trailing behind with their luggage, there were two guards in black Panthera uniforms and similarly colored boots stationed at the entrance of the spacecruiser *Mycenae*.

"Greetings, Ms. and Mr. Starbrook. I am Captain Theodore Ramsey, and this is Lieutenant Viola Narumbi. We look forward to helping ensure the success and safety of your special assignment."

Tessa, who was dressed in a Squad Commander uniform of blue and black, saluted the two guards, even though she was not commissioned yet in Star Fleet.

"Where is our co-pilot?" asked Tomus, who was clad in a baggy sweater and loose-fitting trousers.

"The Commander is at the helm running the ship through its pre-launch diagnostics," declared Captain Ramsey. "Welcome aboard."

When the twins entered the cabin, they could spy on someone in the captain's chair. The co-pilot was dressed in the flight uniform of a Sidereal Federation Fleet Commander – black helmet, close-fitting blue flight suit with appropriate epaulets and tall black boots. From the confidence and grace with the co-pilot went through the pre-launch protocols, Tessa and Tomus sensed a familiarity that led to a mutual feeling of unease. Sensing their presence, the Commander turned to face them, taking off her helmet and smiling while saying,

"Did you really think you could fool your grandmother?"

Chapter 11.

On the two E-w. journey to the *Gilese 667 C* system (through two hyperknots), Tanita informed the twins how she became aware of their scheme to gain passage to Cognito. Being suspicious of the imperative to visit the planet, Tanita contacted Reg, who found out from Finneas O'Rourke, Tessa's desire to investigate crimes across Human Space. Realizing Tessa and Tomus were looking anew at the murder of Colonel Margiven, Reg and Tanita easily realized that the mission to find a venue for Alyie's birthday getaway gift was a disguise for the trip's intended purpose. Reg was initially going to prohibit the journey, but Tanita convinced him that it should go as planned with her as their chaperone.

"Reg and I are not the only ones aware of your devious scheme. Your mother now knows about it."

"She does?" expressed a quizzical Tomus. "How come she is letting us do this?"

"It took some convincing on my part," said Tanita, "but Alyie has always suspected the events on Cognito were more than the authorities were willing to admit. Your interest, in this case, gives the Federation an opening to find out the truth without violating the Act of Self-Determination. After all, who would object to the mother and children of the High Governor looking for a suitable site to celebrate her 50th birthday?"

"Yea, Mom!" Tessa exclaimed.

"But you both will have to abide by my rules, understand."

"Yes, Grandmother."

Chapter 12.

At their arrival at the Hallows Inn, Tanita, Tessa, and Tomus were treated as old-fashioned royalty as befits the family of the ruler of the Federation. The Inn's hotelier, Asher Weinbein, provided a warm welcome. He was a stately older gentleman of erect posture, snow-white hair, piercing blue eyes, and striking good looks. He was dressed in a black fitted suit over a black pullover sweater. His black shoes shone brightly in the warm daylight. Bowing gallantly to Tanita, he declared,

"Madam Commander, what a pleasure to have you and your brilliant grandchildren as our guests. I will personally ensure that your stay at the Hallows Inn is all you have imagined it could be."

"My grandson Tomus has informed you of the intent of our visit. We hope that our inquiry into the suitability of the Inn as the venue for the event we are planning will not be encumbered by any of your staff, and we will be free to interview them. We also rely on your discretion to see that our presence receives minimal outside publicity and its purpose not be disclosed."

"Exactly as you wish, Madame. The whole staff of the Hallows Inn is here to serve you. I have abided by your request to have Nina Mori be your's and your granddaughter's personal attendant, and Carlyle Denton will serve the same role as Master Starbrook. They are both excellent in their service. May I ask how you became aware of them?"

"We have received very good references from some of your prior guests," interjected Tomus.

"I see. I am relatively new here at the Inn, only recently appointed to this position. Enough about me; let me personally escort you to your chalets."

Trailing behind Asher Weinbein, Tanita, Tomus, and Tessa were the two Panthera guards, announced as regulation Federation security, toting the visitors' baggage. Tanita and Tessa were directed by Weinbein into one of the chalets on a radial path away from the main house, while Tomus was sent to an adjacent chalet. (The Panthera would be housed together in a building off the circumference of the Inn.)

Waiting at the door of their chalet, Tanita and Tessa were greeted by a smiling Nina in yellow and white pants and a pullover outfit, who welcomed them and gave them a quick tour of the chalet. Tessa, who noted that Nina was approximately the same age as she, felt there might be an opportunity to extract some information from her.

After showing them the amenities of their lodging, Nina declared, "If there is anything either of you need, don't hesitate to contact me. I am at your service during your entire stay."

"I would like a tour of the grounds while there is still daylight if it is not too much trouble," Tessa chirped. "I need to freshen up and change. Could we do this in about 2 E-hr.?"

"I would be delighted to show both you and Madam Commander the entire property."

"My Grandmother will not be joining us; the long journey has fatigued her."

"As you wish, Mademoiselle Starbrook."

"Nina, please call me Tessa."

Chapter 13.

Tessa changed out of her uniform into casual clothes of a short-sleeved light green tee shirt and cotton-dyed blue pants. She wore durable shoes with treads as she anticipated some hiking over the Inn's property. Wanting to gain Nina's trust, Tessa contacted the guards to refrain from following her. At the precise scheduled time, Nina announced her presence outside the chalet. She was also dressed comfortably, and the two young women looked very much a pair.

Instead of walking around the estate, Nina had secured a ground vehicle which they both hopped in, with Nina doing the driving. They did a quick tour of the stable area, went around the other side of the Inn to the lake, and then headed across the meadow. As they rode together, the two girls shared respective stories of growing up, although their histories were drastically different, with Nina's confined to the locale around the Inn while Tessa was a child of Human Space. Tessa could see that her tales of Kanva, New Cedonia, Teegarden, and Earth aroused in Nina a longing and desire to live a more exciting life.

"How long have you worked at the Inn?" asked Tessa after Nina had stopped the vehicle at the top of a hill to give the former a panoramic view of the valley.

"It is now about two E-yr. I started just after my 16 E-yr. birthday," replied Nina.

"You must have seen a lot of guests come and go."

"Yes, we have had quite a distinguished group of visitors."

"Are any more memorable than others?"

"No, we are required to treat them all the same no matter what their background."

"I meant Nina, did any of the visits result in unusual events?"

"We are pledged to respect our guests' privacy."

"In looking at a venue for my Mom's birthday gift, I read there was a murder here about one year ago. Do you recall it?"

"Oh, Mademoiselle Tessa. Yes, someone was killed in the woods by the lake after an attempted robbery."

"Did you know the victim?"

"No, I did not. It was a gentleman, I believe. My preference is to serve ladies."

"Yes, and I am happy about that. But I understand the murder victim was with a lady when the robbery occurred. Did you know her?"

"Yes, I did. I was her personal attendant but was not in the room when the scarf was stolen."

"Do you have any thoughts on who did this?"

"No, Mademoiselle, the area around the Inn is very peaceful, and we have had almost no crime over the years."

"I hope the area is not too peaceful. I was looking to have some fun and pleasure while I am here. Perhaps, you and I can partake of it together."

"Mademoiselle, I mean Tessa, I would like that. There is something, but it is pretty mysterious. It is a group I belong to. We have meetings in the woods at night, and we chant, dance, and do all kinds of things around a fire after we drink a consecrated wine given to us by our leader that produces a wonderful effect. It is hard to describe; you have to experience it yourself. It's as if another being is inhabiting you. You feel so powerful, as if you are superhuman. I believe I am free to do anything."

"Nina, that is exactly what I am looking for! I am so tired of being my mother's daughter. I would like to break free from my role and become someone else. Can I join you next time your group gets together?"

"The meetings are only open to members, but I can present you as a guest who is thinking of joining."

"Yes, I would like to do that. Just think how great it will be for us to do this together," exclaimed Tessa as she squeezed Nina's hand. "What is the name of your group?"

"The Bacchante. I think it comes from ancient Earth history."

Chapter 14.

While Tessa was traveling around with Nina, Tanita, and Tomus decided to take a stroll to the lake via the glade for the express purpose of examining the area where Colonel Margiven was attacked. (The Panthera stood as sentries at the periphery of the glade.) Tanita was dressed in an outfit of a black faux leather jacket, under which she wore a crisp white sleeveless pullover shirt. Tight violet pants, a silver belt, sturdy black boots, and a black silk scarf around her neck gave Tanita a very sleek look. On the other hand, Tomus was clad in a brown and red checked buttoned-down shirt, brown corduroy pants, and hiking boots.

Starting from the location of Alison's chalet, they walked into the glade in the direction Margiven pursued the individual who stole Alison's scarf. Of course, since an E-yr. had passed, there was scant evidence of a prior struggle. Some bushes were shorter than others, and there was a barren area indicating the site where the attack occurred. However, after a search through the dense forest, Tomus found some vines that led to a clearing some 200 m from the presumed assault. At the center was a flattened boulder, in front of which was a circular mound where there was evidence of a recent fire. Calling his grandmother to join him, Tomus and Tanita inspected the area and found in the embers, bones, and some partially burned pendants. Collecting the objects, they marched back to the main house of the Inn, where Tanita contacted Asher Weinbein.

Appearing presently, Mr. Weinbein greeted them warmly, saying, "Madam Commander, I see you have made yourself at home at our Inn. How striking you look! And Master Starbrook, it is good to see you are dressed for some outdoor activity."

"Mr. Weinbein, my grandson, and I were strolling through the woods near the lake and found these articles in a pile of burnt wood and coal that we believe was a pyre. Do you know anything about this?"

Taking a close look at the retrieved bones and amulets, Weinbein showed no recognition.

"The bones appear to be from some animal, and the amulets look like copies of ancient emblems, but I am hardly an expert. Let me contact Carlyle Denton, who has an amazing knowledge of arcane artifacts."

After a short wait, Carlyle appeared dressed in a tan-colored sport jacket, white dress shirt, dark brown tie, and similarly colored dress pants. His faux leather brown shoes looked recently shined. A medium-built individual of Tomus' age, Carlyle was light-skinned with sandy blond hair and light brown eyes.

Bowing slightly to Asher Weinbein, Tanita, and Tomus, Carlyle asked, "Sir, you called?"

"Carlyle, what do you make of these trinkets? They were found in the glade overlooking Lake Cithaeron by our guests."

Taking a circular-shaped pendant in his hand, Carlyle looked pensive and, after some time, spoke, "It looks like some synthetic jewelry, probably discarded by its user. We have had reports of some activity in the woods surrounding the lake recently. I am sure it is some harmless fun by some of the younger folks living in this area."

"Thank you, Carlyle", responded Weinbein. "So you see, Madame Starbright and Master Starbrook, it is really of no concern."

As they walked back to their respective chalets, Tomus whispered to Tanita, "Even I could see the pendant Carlyle was examining had a grapevine inscribed in its center. It is clearly a symbol for the god Dionysius. Either Carlyle is not the expert he claims, or he is part of whatever is going on here."

"Tomus, Carlyle is your man servant. If what you say is true about him not being forthright about the pendant, then you need to be very careful around him. Do you want me to assign a Panthera to stay with you in your chalet?"

"No, Grandmother. I can take care of myself, and besides, I may be able to get some information out of Mr. Denton if we are alone."

Chapter 15.

When the planetary system's primary star, Gilese 667 C, snuck below the horizon, the Starbright-Starbrook family gathered for dinner in the Grey Gander, the Hallows Inn's main dining facility. Asher Weinbein, in formal attire, personally supervised the autobot staff in serving a sumptuous meal of roast squab, leeks, and whole grain brown rice that was complemented by some locally bottled red wine. Throughout the dinner, Asher was very attentive to Tanita.

When they were finishing a dessert of ice cream with a créme anglaise topping, Tessa whispered to Tanita. "I think he likes you."

"Who?" was her grandmother's reply.

"Asher, of course. Haven't you noticed when he is with us, he can't take his eyes off you?"

"This is not part of my life anymore, and besides, he is not my type. Too genteel."

"Oh, Grandmother," said Tomus. "Loosen up a little."

"You are a very attractive woman," Tessa declared. "I am not surprised a man like Asher would come on to you."

"Aren't we here for more important reasons than flirtations with the hotel staff?" snapped Tanita, wishing to put an end to the topic. "Let's share what we have learned during the course of the day."

Tessa spoke first. "Nina Mori has let me know her Bacchante group is gathering tomorrow after dark in the glade. I am going with her."

"Do you think that is wise?" Tanita asked.

"I will be introduced as a guest, interested in learning about the group's practices. I gather the wine they drink has something put in it that produces in them what Nina calls the Esoteric State. Once this state is achieved, Nina claims you lose your sense of self, and an ancient spirit enters into you."

"This fits with the trinkets we found in the woods. These rites are ceremonies evoking Dionysius," exclaimed Tomus. "Be careful, Tessa; if you pretend to be an initiate, they will want you to drink with them."

"Tomus, I know. I will need to find a way around imbibing the wine."

"This all sounds way too dangerous," replied Tanita. "How do you know this Nina is not trying to lure you into being this cult's next sacrificial victim?"

"Well, I am going to spend the day with her tomorrow, so I should be able to get a better sense of her purpose. I think she likes me and wants me to join the group."

"Does she like you enough to tell you what she knows about Colonel Margiven's murder?" asked Tomus.

"Not so far," replied Tessa

"Looking at the information that's been collected, we still don't have a clear connection between the activities of this Bacchante cult and the attack," concluded Tomus. "Everything is pointing in this direction, but, as they say, we need direct evidence. You are taking a big risk, Tessa, but your presence at this rite may give us all we need to solve this mystery."

"I have a lot of trepidation about this, Tessa, but if you are going to offer yourself as bait, Tomus, I, and the Panthera need to be close by in the glade. Keep your implants activated and signal us at the very first sign you sense danger."

"Don't worry, Grands; I will."

Chapter 16.

For their outing together the next day, Tessa told Nina they would reverse their roles, with the former now taking the latter on a tour. Motoring to the spaceport, Tessa invited Nina to take a ride with her on the spacecruiser *Mycenae*. Seeing the spacecraft, Nina was hesitant.

"Don't be afraid; it is perfectly safe. I am a registered pilot."

"It is just that our group shuns space flight. We believe it is better to travel in the mind than in space."

"Well, I don't plan on taking us into deep outer space. Just for an orbit or two around Cognito. You do want to see the planet you live on, don't you?"

"Ok, I trust you," Nina responded.

After getting on board, Tessa outfitted Nina in one of her spare flight uniforms while she donned her own. Since the two women were approximately the same size, the suit fit well on the Cogitian.

Lifting off quickly once the spacecraft was powered up, Tessa executed a gentle orbit insertion maneuver once a sufficient altitude was reached. Now, in free fall around the planet, Tessa could see by Nina's expression the thrill and wonder of experiencing spaceflight for the first time.

"Wow, this is incredible," Nina exclaimed. "I can't believe we are in space. Everything is so beautiful. Even the darkness."

"Nina, it's great you feel that way, but remember, Cognito is just a tiny part of everything. There is a whole Universe of beauty and wonder to see."

"But I am not the High Governor's daughter who can travel anywhere she wants."

“You are correct, Nina, and because of my position, I can invite anyone I want to join me in my travels. Where would you like to go?”

“Today? I would lose my job.”

“No, when you are available. Pick a place – Earth – Luna – Mars – Kanva – Ergo, anywhere in Human Space, and I will make it happen.”

“You make it sound so good, but I am committed to the Bacchante, and we are against anything that has human technology. We believe we need to abandon human devices and put all our faith in our Lord.”

“Do these ceremonies your group perform help you get close to your lord.”

“Yes, the rites allow us to reach a higher state that allows the god to enter into us. Once there, we become Dionysus and do whatever he wishes.”

“What are the kind of things your god wants you to do?”

“I don’t know. Whenever I reach the Esoteric State, I lose consciousness. I only realize that something happened after I awaken.”

“What do you notice, Nina?” Tessa inquired.

“My clothes are soiled and sometimes torn. My face is dirty, but what bothers me most is my hands and fingernails.”

“What has happened to them?” Tessa pressed.

“The hands have what looks like dried blood on them, and the nails are broken with traces of blood and soft flesh-like material clinging to them. It really creeps me out.”

“I can see that it would, Nina.”

"But trust me, Tessa, as you get near the Esoteric State, you feel so wondrous. Tomorrow night, let's hold hands when we go through it. I want to experience it with you."

"Sure, Nina, we will be together."

Chapter 17.

After nightfall, Nina came to Tessa's chalet to take her to the ceremony. Tessa was dressed in her Academy cadet outfit, but Nina brought along two black robes that they would put over their clothes once they were near the site of the meeting. Tessa could see that Nina also carried two masks similar to the one described by Alison.

When Nina spied a laser pistol that Tessa had tucked into her left boot, she said, "No weapons are allowed in the sacred grove. Our Lord forbids any human devices of destruction."

Reluctantly, Tessa left the weapon in her chalet room, inactivated. She did, however, send a signal via her implant to Tanita and Tomus, who were in the Grey Gander, alerting them she was leaving with Nina.

Walking together, Nina and Tessa entered the glade and followed a path that brought them near the site Tanita and Tomus had discovered yesterday. There was a light coming from inside the enclosure and the sound of voices murmuring or chanting. Nina helped Tessa put the robe over her head while she did the same with hers, which had a distinguishing stripe. Giving Tessa her mask to wear, the women stepped through the vines and entered the grove.

What Tessa realized was that the secluded bough was an outdoor chapel. The boulder described by Tomus served as an altar, and the dirt mound in front, which had a slow-burning fire, was the sanctuary. There must have been at least twenty people chanting in small groups. Nina held tightly to Tessa's hand as they drew closer to the fire waiting for the ceremony to begin. When a robed figure with a stone-like mask under a monk's cowl entered the grove holding what looked like a thyrsus, Nina whispered to Tessa, "That is our leader; it will start now."

This person shod in wooden sandals passed through the assemblage up three stone-like steps to the height of the altar and, holding up the staff, in a booming voice, proclaimed,

"Bacchante, welcome to our sacred mysteries. On this night, we will enable Dionysus to come into us and direct our actions. We will be free of thoughts and fears but allow our Lord to unleash our true natures. We renounce this world and all its trappings and allow our godlike nature to come forth."

As the leader's declaration was met by assenting nods, goblets were passed out, and a barrel of wine was brought up to the leader's lectern. Taking a vial containing some liquid from inside his/her robe, the leader poured its contents into the vat of wine, saying,

"Elixir of Dionysus, enter the hearts and minds of your disciples here before you. Fill them with your spirit and grant them the power to shake off their limited human reason and become a god like you."

As the goblets were filled with the wine and passed among those present, the leader indicated with a gesture not to drink yet.

"Brethren, I have exciting news tonight. We were to present a goat to our Lord, but it has changed. We can now give him a more deserving gift. We need to thank our sister Nina for bringing the offering."

As all eyes turned towards Nina and Tessa when the leader pointed them out. Nina appeared surprised at this pronouncement. Before she and Tessa could react, hands pried them apart. Tessa was taken hold of and unmasked with her robe stripped off.

Nina protested, "What are you doing? She is my guest and wants to learn about our group. Let her go."

"Oh, Sister," scolded the leader. "You are too easily deceived. She and her people are trying to find out what happened a year ago when that infidel was sacrificed to our Lord. I was suspicious since their arrival here. When they started looking in the glade and asking me questions, I knew their real intentions."

"No, that's not true. She is my friend," cried Nina.

"She tried to use you to find out about our practices, but our Lord had other plans and made you the conduit to bring this offering to us. What better sacrifice to our Lord than the daughter of the leader of the godless Federation? Sister Nina, we praise and thank you."

Tessa was dragged up the steps by four hooded acolytes. During her struggle, she tried to access her implant but found it could not be activated. Apparently, the grove had some security shield surrounding it. Her jacket and top were ripped off, and she was splayed down on the altar by each acolyte holding either her hands or boots. All the while, the rest of the group was guzzling the wine and chanting louder and more uncontrolled.

"Once Dionysus returns, the consciences will flourish. My child, bring me a grail. I think I have the singing abilities of Dionysus. Now, child, bring me a grail. I've been left with a crown of ivy, and I crash it all upon my anger. My child, bring me a grail. Feed me, although I drink. Now, child, bring me a grail. Now, child, bring me a grail."

As the group was about to reach *bakkheia*, the leader raised the thyrsus high to signal a halt in the incantation. In the leader's right hand, a long curved dagger was seen dangling above Tessa.

Tessa, stretched painfully by all four limbs, managed to blurt out, "You won't get away with this. Others know why I am here."

"The Lord is directing me against your Federation, which has tried to take the mystery out of the Universe. He will keep me safe," was the leader's reply with the dagger poised above.

"Bacchante, I will now pierce her heart. After I draw first blood, I give her to you to tear apart."

Just as the dagger was about to plunge into Tessa, Nina rushed up the steps with a club. She struck the first acolyte who tried to impede her but fell down when the others rushed, preventing her from freeing Tessa. Falling down the steps, she was attacked by the mob that was now in a manic state and savagely beaten. While she was being brutalized, the leader pronounced, "Sister Nina has been stained by her contact with the infidel. Bacchante. Let the Lord use you as the instrument of her punishment."

Tessa was again restrained on the rock. Just as the leader was poised to stab her, shouts were heard in the distance. Turning to face the source of the noise, the leader was struck by a laser blast right in the face, setting most of it on fire. Other flashes of laser beams were seen as Asher Weinbein, toting a laser rifle, and the Panthera guards, Captain Ramsey, and Lieutenant Narumbi torched the acolytes holding Tessa.

While the mob below was being subdued, Tanita and Tomus rushed up the steps to Tessa. She was shaken but unscathed. Tanita found a robe to cover her granddaughter's upper body. Even though his head was badly burnt, Tomus could recognize enough of the corpus to see that the leader had been Carlyle Denton.

Tessa's first thought after being assisted down the steps by Tanita was for Nina. "She tried to save me."

At the foot of the altar near the pyre, they found Nina's mangled body, beaten terribly by the frenzied mob. Tessa sobbed uncontrollably in her grandmother's arms.

Coming near was Asher Weinbein, looking very comfortable in a Panthera uniform. When Tanita sighted him, he nodded. "Madame, the area is now secured."

Coda

The *Mycenae* had just come out of the hyperknot that put it in the Rebus Centauri system. Tessa engaged the spacecraft's ramscoop to accelerate in normal space towards New Cedonia. Next to her in the co-pilot's seat was Captain Theodore Ramsey. As Tessa monitored the hydrogen fuel infusion, a request for a video hyperlink transmission was received. Granting it, Tessa gave the ship's command to Ramsey and went to the cabin where Tomus was waiting to join in this hyperlink meeting as Lieutenant Viola Narumbi took over in the co-pilot's seat.

On the twin's screens appeared the anxious but somewhat relieved faces of their parents, Alyie Starstriker and Jaynis Bottombrook.

"Hi, Mom and Dad. I hope you are both well," chimed the twins.

"Tessa and Tomus," replied Jaynis, "both of you are causing your parents to age faster than normal."

"But we solved the mystery," said Tessa, hoping to deflect her parents' irritation.

"But not before you were nearly sacrificed to a 4000 E-yr. old deity" was Alyie's swift rebuke.

"Who knew that bough had a security shield?" was Tessa's answer. "Everything would have gone to plan if I just could have activated my implant."

"It was great that Asher Weinbein found out about our plan from Captain Ramsey," said Tomus. "Who would have guessed Mr. Weinbein had served in the Panthera. Apparently, Reg Jones had alerted him of the real purpose of our visit. Once he realized that Tessa was in the hidden glade, Mr. Weinbein knew she would not be able to contact us."

"After the discovery of this cult in Cognito," said Alyie, "we have identified other cells of this sect in Ergo, Kanva, and Nespar that local officials are disbanding as we speak. The drug taken, lysergic acid diethylamide, to produce the hallucinatory state has been identified, and its source of production has been located. All quantities have been confiscated. So, in total, your efforts were a success. I just wish you weren't so reckless."

"Do we know why Carlyle Denton picked out Colonel Margiven and Alison as intended victims?" wondered Tomus.

Jaynis offered this explanation. "As far as the authorities could piece out, the Colonel was chosen on a whim after he had somehow offended Denton, who was his manservant at the Inn. Of course, if Alison had joined in the pursuit of her scarf, she too would have been offered as a sacrifice to Dionysus. It is inconceivable to think in the 27th E-c., there are still pockets of humanity who believe in these old myths and blood rituals. I thought we were beyond superstition and human sacrifice."

"Changing the subject," Alyie interjected, "where is my mother? My implant can't locate her on the spacecraft."

"Oh, Grands stayed behind in Cognito," answered Tessa. "Mr. Weinbein has promised to show her all the delights and pleasures of the planet. She, in turn, wants to take him up in space so he can experience love in 0 g. She changed her estimation of him once she saw how handy he was with a laser rifle. They make a cute couple."

"My goodness," exclaimed Alyie while turning to Jaynis," Cognito must be the place to unleash the animal within. Maybe we should go there for my birthday?"

"Now that we are past the terror of what might have happened," Jaynis interjected, "I understand our daughter has set a record."

"Yes, I am the youngest person to take a Federation spacecraft through a hyperknot," beamed Tessa. "Who was the previous record holder? Let's see. Her name is on the tip of my tongue. Do you know, Mother?"

"Yes, I do, but remember when I accomplished this at 21, I was under an official Federation command."

"Touché, Mom."

---

*From *Assassin's Creed Odyssey*, Ubisoft, 2018

The RIPEST Harvest

It is wrong to think the task of physics is to find out what Nature is.
Physics concerns what we say Nature is.

Neils Bohr

Chapter 1.

The aircar deposited Alison Bernewithy and her guest Yong Gwan at the Bradbury Hotel, where she was staying during her official visit to Mars. In the fourth E-yr. of her term as Regional Governor of the Sol System, Alison was conferring with scientists at Chandrasekhar University as part of her regular purview to identify the future needs of this research community. Presently she was returning from a gala dinner and reception honoring some of the most distinguished faculty at the university. Dr. Gwan received the Percival Lowell Award for Early Career Achievement for his work in developing an algorithm that ascertains the suitability for humans habiting exoplanets based on signature trace gases. After the awards ceremony, Alison, in conversation with Dr. Gwan, indicated that she wanted to know more about his work and invited him to join her afterward in her suite for what, at an earlier age, was called a nightcap.

Stepping into the lobby of the Bradbury in her tall black high heels, Alison removed the thin capillary tubes from her nostrils, breathing the ambient oxygenated air freely. She was clothed in a full-length, thermally insulated black satin coat. Directing Dr. Gwan into the lift, Alison's voice commanded it to take them to her suite on the 10th floor. Stepping out of the lift, Alison and Yong walked down the corridor to her room which she bade open. Once inside, she had Yong assist her in removing her coat, revealing her stunning white dress. It was contoured to hide the excess weight Alison now carried but accentuated her lovely neck with a plunge that descended to the top of her breasts. After removing her mid-length black gloves, she straightened the Regional Governor insignia that was pinned near her left breast. Her colored yellow hair was pinned back with a gold clasp, and her blue eyes looked more striking through the use of eyeliner.

When Dr. Gwan removed his heavy overcoat, he was dressed in a formal brown suit. He wore under his jacket a brown turtleneck top, and his sturdy brown shoes had treads underneath. In his mid-20 Eyrs., Yong Gwan was a pale-skinned, black-haired young man of mid-height and built with dark piercing eyes and high cheekbones.

"I am going to have a drink; can I get you something?" Alison asked as she directed Yong to sit on the sofa in the anteroom of her suite.

"What is available?" responded Yong, unsure of what was the proper protocol.

"Let's see what the device in my room can produce," said Alison. "I see Scotch whiskey, Cognac, and wine, all made, I am sure, from formulas concocted in labs here on Mars. We can also ask management to provide other options if these don't suffice."

"What will you have, Madam High Governor?"

"I think I will have some Cognac as I still feel a chill from the Martian air, and please call me Alison."

"Then it's Cognac for me also."

Alison dialed in that selection and a few E-min. later poured the emerging liqueur into the goblets provided. Raising her glass and toasting, "Here's to your research and a new day for space exploration."

After they imbibed the elixir, Alison continued. "If I am correct in understanding, your algorithm will dispense with the task of sending drone spacecraft on missions to find habitable worlds."

"Yes, Madame, I mean Alison, with the Remote Identification of Plausibility Environments (RIPE), just using minimal data, we can determine whether a newly discovered planet is suitable for terraforming into a habitable environment."

"This will save time and resources for the Federation," Alison surmised. "We are talking about many E-yr. difference in beginning human settlement. I need you to accompany me to Federation Headquarters on Earth to present this to High Governor Starstriker.

"Really? I haven't been on Earth in some 10 E-yr. It will be good to see it again."

"Well, you're in for a treat. Before we go to HQ in Colorado Springs, I will take you to the Regional HQ Office in Brussels, where you will meet my staff. I will then make time for a quick visit to Paris, where we can stay at my pied-à-terre, and you can experience how vibrant this city has become."

"I do not know what to say, Alison. Thank you for all this."

"How about saying, 'Cheers' to our collaboration."

"Yes, I can drink to that."

"I was hoping, Yong, that we would do more than just toast," said a smiling Alison.

Getting the cue, the young man put his glass down as Alison leaned toward him, wrapping her left hand gently around his neck. Kissing him firmly on the mouth, Alison carefully put her goblet on the side table as she and Yong embraced. He had never been with a mature woman before, and Yong was impressed with how adept she was at synchronizing her arousal with his.

Chapter 2.

Yong's trip to Earth was a dust devil. He was granted leave from Chandrasekhar University to present RIPE to the High Governor and her Board. Traveling with Alison on her spacecruiser *Felicity*, they made the one E-w. trek to Earth, landing in Brussels. At Regional Headquarters, Alison introduced him at a Regional Council meeting as "The brightest scientific light in the Sidereal Federation".

Taking the Regional Governor's personal airjet to Paris, they encamped at Alison's townhouse, where Tomus Starbrook was also residing. Now almost 20 E-yr., his hair color split down the middle in blue and green, and his skin tone a bright yellow. Tomus viewed the young scientist skeptically.

Alison tried to defuse the tension, saying, "Tomus, this is Yong Gwan. I am bringing him to your parents so he can present his new algorithm for space exploration."

"I am pleased to meet you Dr. Gwan. I hope your stay on Earth finds success for you."

"Thanks, Mr. Starbrook."

"Tomus is an archivist at the Louvre Museum. Can you arrange for Yong and me to have a personal tour?"

"Alison, you don't need me to do that," was Tomus' unenthusiastic reply.

"Yes, but Tommy, you are so good and have such knowledge of the Museum's collection. I want it to be special for Yong."

"As you wish. When would you like to visit?"

"Give me a day to outfit Yong properly for his stay on Earth. Let's make the tour two E-d. from now."

"He seems dressed okay to me."

"Yes, but not for the places I want to take him nor to meet your mother. Tomorrow, we are going to Saville Row."

"London? Why do you need to go there? You can just use your 3D fabricator to clothe him."

"Oh, Tommy. You are so fashion challenged. I want Yong to have bespoke clothing, namely the experience of having a human tailor fit him perfectly. How is that by you, Yong?"

"Whatever you think will help me, Alison, I will do."

"Then it's settled. London, tomorrow. Paris the next."

Chapter 3.

Early the next E-d., Alison and Yong, after a hearty breakfast of croissants (with real butter), strawberries, crème, and hot drinks, ascended up the flight of stairs to the roof of Alison's townhouse where the airjet was waiting. Alison was dressed smartly in a yellow suit consisting of a short cropped jacket and pencil skirt over which she draped a camel-colored double-buttoned trench coat. Underneath the jacket was a white blouse knotted with a flowing white silk tie. She wore tall high-heeled synthetic leather black boots and her signature black gloves. Yong was dressed in a brown corduroy jacket, open-collared white shirt, casual brown pants, and sturdy brown walking shoes. After Alison announced to the airjet's control system the intended destination, they were airborne and, in less than 30 E-min. were put down on the roof of a building in Mayfair.

Taking the lift to the street level, Alison and Yong walked the short distance to the shop on Saville Row through a light London rain, where she and Yong had an appointment for his fitting. Noting the entrance of the Regional Governor, the head tailor rushed to Alison, bowing while kissing her gloved hand. The shop swung into activity, measuring Yong while Alison looked at clothes and chose what styles of jackets, trousers, shirts, and ties she wanted for Yong. After an exhausting four E-hr., Yong was now the somewhat bewildered owner of four bespoke suits, five trousers (including breeches), 10 starched cotton shirts, and 20 silk ties. A shoe designer was brought in and fitted the young man with a pair of black and brown dress shoes and black riding boots.

Dressing fully in one of his new suits (while all other purchases were sent to Alison's pied-à-terre), Yong presented himself in front of Alison. Beaming, she declared, "Now you look like a gentleman. Don't you think you should take a lady to High Tea?"

A frazzled Yong could only say, “Yes, Alison, if that is what you want.”

Stroking his hair gently with her right glove, she kissed him lightly on the mouth, saying, “I know just the place.”

Chapter 4.

Lifting off after the stay in London was concluded, Alison commanded the airjet to circle so that Yong could get a panoramic view of the historic city. As the craft was hovering in the night sky above the Thames River, Alison reached over Yong to identify Westminster Cathedral and the House of Parliament. While he was looking down at these landmarks, he felt pressure below, and turning, he noticed Alison smiling as she was rubbing her right gloved hand beneath his groin.

"I thought we might have a little fun at this height," she cooed. Not knowing how to respond, Yong let Alison get on top of him, straddling her tall boots on either side like she was riding a horse. She started kissing him passionately on the mouth, face, and neck while lifting her skirt and pulling down her undergarment. He reciprocated, and soon after, they were enjoined, rocking back and forth rhythmically as one. They stayed in synch until neither could hold back any longer. Alison groaned in delight while Yong felt satiated.

"I am a bad girl, aren't I? Here I buy you a beautiful outfit, and then a short time later, I cause a mess. But just entre toi et moi, I think it was worth it, don't you?"

"Of course, Alison."

Chapter 5.

They were back in Paris by late evening. The next E-d. Alison gavc Yong a guided field trip to as much of the city's cultural treasures as could be accomplished in a single day. Starting out in the morning, they both were clad in casual attire. Yong wore one of the shirts and trousers he acquired yesterday, but he kept on his old walking shoes. Alison dressed in a houndstooth stitched brown and black blazer jacket worn over an off-white pullover top. She squeezed into light brown tight trousers that were tucked into black riding boots. Her blonde hair was pinned underneath a black beret.

Stepping into a hovercraft, Alison and Yong went first to the summit of Montmartre, where the Sacré-Cœur Basilica was seated. From the height of the hill, Yong gained a sweeping vista of Paris with its winding streets, churches, and historic institutions. Getting back into the hovercraft, Alison next commanded it to set down on the tiny Île de la Cité in the middle of the Seine River where Sainte-Chapelle and Notre-Dame de Paris are seated. The former was a chapel for the Kings of France from the 12th to 14th E-c. (CE), and the latter a cathedral first constructed in the 12th E-c. Yong, who had little education in Earth's early history, was amazed at the craftsmanship and artistic beauty in the architecture of these hallowed places. Next, they traveled to the Macron Galerie, where the most contemporary art was displayed.

Mid-day, Alison escorted Yong to the Louvre, where Tomus was dutifully waiting to greet them.

"Welcome to the treasure house of Earth culture," Tomus pronounced. "What would you like to see?"

"Oh, Tommy, you have such good taste. Give us a sampling of what a neophyte like Yong should see on his first visit," Alison interjected.

With that guideline, Tomus took them on an idiosyncratic excursion into the labyrinth of the Museum's galleries, showing both familiar and arcane works in the vast collection. In the few E-hr. the three of them toured, they viewed a range of Earth's artistry from early times to almost the present. Exhausted both in mind and body (especially in their legs), Alison and Yong thanked Tomus for giving them his time and astute insights into the art they had viewed.

Retreating to the apartment to recoup from the day's events, Alison and Yong, once refreshed, dressed for the evening in which Alison had arranged to be a formal dinner and attendance at an opera performance. At the *Epicure* on the Rue du Faubourg Saint-Honoré, they dined on quail, red potatoes, and leeks served with champagne from harvested grapes. Alison insisted that Yong wear one of the suits (black) she had purchased for him while she wore an almost figure-bursting green gown with elbow-length white gloves and very high black heels. Her hair was pulled tight with a diamond gold pin.

"This is all overwhelming," Yong asserted as he sipped his champagne while finishing dinner with a dessert of gateau au chocolat. "I don't mean to sound ungrateful for all you are doing, but I have to tell you, Alison, much of it may be wasted on someone like me. As a scientist, I don't need to know all this culture or wear formal dress. People at my lab would laugh if they saw me like this."

"Oh, my dear boy," Alison replied, "what I am doing for you is not intended for your lab mates. If you want to be influential in the Sidereal Federation, you need to have a sense of history, culture, and panache. High Governor Starstriker also puts a high value on personal appearance. Think of these clothes as a uniform. Don't worry; it is paid for in my budget. Trust me; I know what is needed."

"But, I am not sure I want to be an 'insider'. I am very happy working on Mars at Chandrasekhar."

"But I am not. Yong, I know you are perfectly satisfied with your achievement, but in a career, you cannot rest on a singular success.

Seeds need to be planted now to ensure your future. I know that Dr. Jaynis Bottombrook, Chief Scientist for the Federation, is thinking of stepping down from his position when his life partner, Alyie Starstriker, finishes her term as High Governor in one E-yr. I am sure Jaynis will be looking to groom his possible successor. I want to introduce you to Jaynis and have him get to know and observe firsthand the potential I see in you."

"This all happening very fast for me. I am not sure I am ready to relocate to Earth."

"Trust me, when the time comes, you will be ready. I will ensure that. Remember, the best part of you working on Earth is that we can be together. You do want that dear boy, don't you?" Alison asked, reaching out across the table to clasp both of Yong's hands in her white-gloved fingers.

"Yes, Alison," was all Yong could say in response.

After their dinner was completed, they were picked up in a hovercraft and whisked to the Palais Garnier, the venerable Paris Opera house, for a performance of the early 20th E-c (CE) work *Der Rosenkavalier* by Richard Strauss.

Chapter 6.

Sleeping in until Tomus left the next morning for work, Alison went into Yong's bedroom and cuddled up alongside him, waiting until he awoke. Opening his eyes, he was startled to see her next to him. She moved to caress him, but he reminded her they had to make haste to catch the aircruiser that would take them to Federation Earth Headquarters in Colorado Springs. After having a quick breakfast prepared and served by Alison's autobot, they washed and dressed. An aircar took them to the Paris spaceport, where Alison had a private hypersonic aircruiser set to fly them to Federation Earth HQ. Cruising at Mach 5.0, the plane made the trip in less than 1.5 E-hr. Disembarking, they were met by a Federation aircar. Yong was dressed in one of his new suits (blue), wearing a stiff white buttoned-down shirt, blue and red striped tie (which Alison assisted in knotting), and black polished shoes. Alison was in a bright red blazer and pencil black skirt, draping over her shoulders a ¾ length leather jacket to provide some warmth from the bracing mountain air. Her white satin blouse was fastened at the top by a looping thin black tie. She clopped alongside Yong in her black high-heeled boots.

Descending down into the spectacular snow-covered canyon that served as the backdrop for the complex that comprised the Federation Headquarters on Earth, Alison pointed out to Yong the recent history of the construction of this center. Ushered into the main building of the complex, both Yong and Alison were allowed to freshen themselves in a lounge until they were called to join the High Governor's Board Meeting.

After waiting for about 30 E-min, they were sent up the lift to the sixth floor, where the Conference Room was located. Entering, Alison could see familiar faces in Alyie Starstriker, striking as ever, dressed in her formal High Governor attire of white jacket, pants, and tall black boots, and her husband Janis Bottombrook in more casual clothes. Also in attendance were some Federation officials who

Alison assumed were scientific and engineering colleagues of Jaynis. On the displays surrounding the room was the imposing face of Vice High Governor Reg Jones, the Military and Security Head of the Federation. His presence assured Alison of the seriousness that Alyie was taking Yong's algorithm.

Leaving the head of the conference table, Alyie approached Alison and hugged her, kissing her on both cheeks.

"How good it is to see you. It amazes me that we reside on the same planet, but it takes a breakthrough like Dr. Gwan's theorem to bring us together."

"But you will see that RIPE is truly remarkable, and it has been worth the wait for me to see you."

Motioning all present to sit, Alyie opened the meeting by introducing Yong.

"Dr. Yong Gwan is a planetary scientist at Chandrasekhar University on Mars. At the request of Regional Governor Alison Bernewithy, he is here today. Alison, will you tell us the reason you have asked for us to meet Dr. Gwan?"

"Thank you, Madam High Governor and the Board, for allocating time for Dr. Gwan and me to be present today at Federation Headquarters. I came to learn about Dr. Gwan's work through my periodic review of the scientific and engineering reports from Chandrasekhar. His development of the Remote Identification of Plausibility Environments, (acronym RIPE), piqued my interest because I know, Madam High Governor, during your tenure, you have placed a high priority on the need for the Federation to expand by finding and developing new worlds for humankind to live and prosper. Learning more about RIPE from talking to faculty at Chandrasekhar, I have recently returned with Dr. Gwan from Mars after attending an awards ceremony in which he received an Early Achievement Award. My understanding, which Dr. Gwan will certainly elaborate on, is that RIPE promises to revolutionize the Federation's ability to identify worlds to settle for human expansion. I hope you will be receptive to his algorithm and see it incorporated in the Federation's future exploration initiatives."

Jaynis rose from his chair and welcomed Dr. Gwan by saying, "We at the Scientific Institute have been impressed at the results your algorithm has demonstrated in an already colonized world. If RIPE can be used to locate new ecosystems suitable for human habitation, we would like to move forward with its implementation. Dr. Gwan, before we question its merits and limits, can you take us through its development?"

Just as Yong was about to begin his presentation, a disturbance was heard outside the Conference Room just to his right, and a person dressed in the uniform of one of the Federation's Elite Guards entered. Approaching the High Governor, he whispered in her ear. Alyie displayed annoyance but eventually nodded affirmatively to the guard. Opening the Conference Room door, the guard ushered in a tall, athletic-built young woman dressed in the off-flight uniform of a Space Fleet junior officer – blue with black trim shirt (with two breast pockets) and pants (with black stripes), black tie and black boots. She had a cap on but, immediately upon entering, removed it, revealing her tousled golden hair streaked with red highlights. Her face had a finely chiseled upward curving nose and symmetrical features, and her sparkling green eyes glimmered in the sunlight room.

Yong, upon seeing her, was dumbstruck and didn't know what he should do. He stared at the High Governor, who responded, "Dr. Gwan and colleagues, excuse the interruption. This is newly commissioned Lieutenant Tessa Starbrook, who has asked to join our meeting. Being an indulgent mother, I have acceded to her request. Please find a seat, Lieutenant, and proceed Dr. Gwan. I assure you there will be no further disturbances."

Chapter 7.

As Yong stepped through his presentation on RIPE, he had difficulty trying not to stare at Tessa, who stationed her chair right in front of where he was speaking. She seemed entranced by his every word leaning forward with her elbows on her knees, and her boots pressed firmly together. Seeing right below a freckled face so pure and cherubic, Yong felt he was communicating directly to her soul. After providing the specifics of the algorithm and how using a data set of physical parameters such as size, density, atmosphere, and orbital properties, Yong asserted RIPE could compute a plausibility factor on whether a candidate world is capable of human development. Using information from already habituated environments, RIPE had a 99% rate of success in identifying these worlds as suitable. The program also was 100% successful in determining, from the data sampled, uninhabitable worlds.

After the presentation was concluded, Jaynis spoke first, declaring, "I am sure Dr. Gwan would be happy to answer any questions."

Dr. Cybele Wainforth, a planetary scientist, asked, "Dr. Gwan, does RIPE assume Earth-like conditions only as the criteria for determining the plausibility index?"

"No, because with terraforming, the Federation has been able to engineer many different environments for human habitation; the program does not normalize the parameters for Earth. Instead, RIPE has correlated conditions on the over 30 worlds the Federation has settled and uses a statistical average to determine if a candidate fits the profile."

"Have you looked at new candidates for human settlement?" asked Terra Engineer Rufus Steinhaus.

"Yes, RIPE has been able to identify four planets and one moon that has exceeded the plausibility threshold for development."

"Is your program able to determine if a candidate world has life forms?" questioned the exobiologist Liu Fan.

"Not directly, Dr. Fan," was Yong's response. "There are standard measures for ascertaining the presence of life such as methane and carbon dioxide emissions, electromagnetic transmission, and bio-chemical identifications that are used in conjunction with RIPE."

"I have a question," Jaynis asked. "Detection of possibly habitable worlds has been very challenging when a star system has an abundance of exozodiacal light – namely dust particles surrounding the system that obscure measurements of the physical properties of the worlds within. Can RIPE surmount this obstruction?"

"Yes, it can, Dr. Bottombrook. RIPE can still give a plausibility rating by using the temperature of spectroscope signature of the exozodiacal light to infer planetary conditions."

Jaynis then noted that Reg Jones, watching remotely, indicated he would like to speak.

"I am very impressed with Dr. Gwan's work and, if it is all he says it is, this can move us forward at considerable savings. The enterprise of sending constellations of robotic spacecraft across the galaxy combing for life-sustaining worlds is kin to prospecting for rare pearls in a gigantic ocean. The benefit-to-cost ratio for these missions is very low."

"As valuable as cost reduction," Jaynis interjected, "is the savings in time. If Dr. Gwan's algorithm can identify target planets and/or moons for development, we can launch the Phase One series of spacecraft with robots and intelligent machines that can begin terraforming immediately."

Alyie interrupted the questioning by asking, "What I would like this Board to recommend is the next step the Federation will take in implementing Dr. Gwan's RIPE."

"Madam High Governor, if I may speak?" Space Systems Engineer Mahmoud Foustak declared as he rose up in his seat.

"I know you place a high economic priority on human space expansion, but we are perhaps being a little too hasty in moving forward with Dr. Gwan's most promising algorithm. I recommend we take a more measured approach, doing a series of simulations with RIPE to see how successful it is on different fabricated environments. This will give us an assessment in terms of its scope and limitations if and when we decide to use it."

Mr. Foustak's recommendation seemed to generate some consensus as heads around the Conference Table nodded in agreement. However, from her front seat, Tessa stood up and turned around, exclaiming, "What are we waiting for? This is a godsend. Let's get a mission put together to see up close if RIPE works. I volunteer to be the pilot."

All looked at her in astonishment at the audacity of her idea. Then eyes turned to Alyie, waiting for Tessa's mother to respond. When she collected herself, she spoke, saying, "Thank you, Lieutenant, for that suggestion. We will take it under advisement. I think we are done here."

Tessa felt embarrassed by the icy response to her proposal, but she took some comfort when looking up at the screen; she noted Reg was trying to suppress a laugh.

As people were filing out of the Conference Room, Alyie and Jaynis joined Alison and Yong in conversation. Jaynis expressed again how impressed he was with Yong's work, and Alyie talked with Alison about Yong's itinerary during his visit to Earth. Hearing that they planned to return to Paris, Alyie objected, "Don't go back immediately. We need to talk more about this algorithm. I am hosting a small informal dinner party at my home for family and some friends to celebrate Tessa's commission. Please join us tonight. I am sure we can find time to discuss Dr. Gwan's work more fully. You two can fly back tomorrow. I will arrange rooms for both of you at the Falcon Inn."

"Alyie, that is delightful. Of course, we will stay."

"Before you leave, let me introduce my impulsive daughter to Dr. Gwan. Having just her received wings, I guess you can tell Tessa is aching for a mission."

Beckoning Tessa to join the foursome, Alyie introduced her to Yong, who had trouble responding to her greeting. Tessa, on the other hand, was not timid, taking Yong by the hand while leaning forward (she was half a head taller than him) and whispering in his ear. "I think what you are doing is fabulous, and I want to be a part of it. We need to develop a plan. Don't worry; I have ideas."

Not knowing how to reply, Yong just bowed his head, mainly to disguise his flushed face. Alison, standing nearby, saw this all and had a melancholy foreboding of what was to come.

Chapter 8.

Sincc both Alison and Yong had not anticipated staying over in Colorado Springs, they had not packed any other clothes, so they were dressed for Alyie's party the same as at the Board Meeting. (Although Yong had shed his tie.) He wanted to arrive separate from Alison, but the High Governor had an airjet waiting to take them to her mountain chalet. Entering the High Governor's mansion, Yong was struck by the high ceilings and expansive rooms. Living on normally sub-freezing Mars, interior spaces were compact in order to maximize heat flow. At the party, Yong could see the High Governor and Dr. Bottombrook. Talking to them was an elegant-looking older woman that Alison informed him was Commander Tanita Starbright, Alyie's mother, who was accompanied by a debonair-looking gentleman, Asher Weinbein, who was now her companion. There were some other guests milling around drinking and eating that Yong assumed were Federation officials, one of which Alison whispered was Retired Admiral Steely Cranshaw, a mentor of Alyie. Tessa could be seen conversing with two younger-aged guests, presumably her peers, holding what looked like a miniature dinosaur in her arms. Everyone was dressed casually; even the High Governor had exchanged her uniform for a low scoop pullover top, Capri length pants, and flats.

Dr. Bottombrook was featuring his home-brewed beer, and the main course consisted of smoked synthetic bison, a take on Rocky Mountain Oysters, and sweet husked corn.

When Alyie spied Alison and Yong entering, she went to greet them, hugging Alison (again) and expressing her admiration for Yong's presentation while shaking his hand.

"Dr. Gwan, you are now the talk of Federation Headquarters here on Earth. Your algorithm has excited us with the possibility of accelerating our need to expand Human Spacc."

"Thank you, Madam High Governor."

Alyie then proceeded to introduce Yong to the other guests, including her mother and Asher.

Tanita smiled broadly, saying, "You have really made quite an impression, especially I understand with my granddaughter."

Not knowing exactly how to respond, Yong was spared as the conversation with Tanita was interrupted. Tessa joined in, holding what looked like a reptile sporting two protruding horns in her arms.

"It's about time you got here; we have a lot to discuss."

She was dressed in a simple pullover long-sleeved white shirt with a green sleeveless vest, blue jeans, and soft-soled shoes. The color of her vest made her eyes even more vivid. Taking Yong by the hand, she walked him across the room to a balcony. As Yong was stepping outside with Tessa, he caught a glimpse of Alison, who had her head slumped down.

The vista from the balcony was magnificent. They were surrounded by white peaks in the bracing fresh air.

"This is Beelzebub; he is a short-horned lizard that my parents keep to rid the house of pests such as harvester ants. Isn't he cute?"

"Cute is not quite the word I would use to describe him. Is he friendly?" expressed a wary Yong.

"As in most creatures, you need to take the time to gain their trust. Do you like the view?"

"It is spectacular," was all Yong could muster to say.

"I know the mountains on Mars are taller, although being shield volcanoes, they are not as pointed as these. My folks call this place Eagle's Nest, but I like to think of it as a cradle."

Summoning up his courage, Yong asked, "What made you come to the Board Meeting?"

"It was to hear you, of course."

"How did you know about my work?"

"My brother, Tomus, told me all about you. You met in Paris."

"He was very kind to give Alison and me a tour of the Louvre yesterday."

"You know he doesn't like you."

"Why? We have only just met."

"It's because of Alison. Tomus is lovesick for her. He has been that way since she started caring for us when we were small children. We think of her as our aunt."

"Your aunt?"

"You don't have anything with her, do you?"

"No", Yong replied. "Hardly."

"Good, because we have to make plans and be ready to move."

"Everything is happening so fast for me," Yong declared. "Two weeks ago, I was working in my lab on Mars, and now I am in the beautiful home of the High Governor whom I just shook hands with and I am talking to her daughter. I mean, no disrespect, but is that really the almighty Alyie Starstriker inside? Am I imagining all this, or has someone slipped me some hallucinogen?"

"It's all real, and the reason is RIPE. Trust me, what you have developed is a game changer. My mother is very interested in implementing this, but you have to convince my father, whose judgment on these matters she trusts implicitly. Once he gives her the okay, Mom will act swiftly. That is her nature."

"What do I have to say to convince him other than what I presented at the meeting?"

"It's not what you need to say but what we have to do. We need to propose a proof of concept mission to convince Dad and the others at the Scientific Institute. By tomorrow, we need to have it all set up."

"How can we do this on such short notice?"

"I have it all laid out. You need to talk to Dad tonight and volunteer to be the lead scientist on an exploratory mission to one of the candidate worlds your algorithm has selected as having high plausibility for human habitation. Think about what other team members you will need to go with you."

"But I am a scientist, not a space explorer."

"Don't worry; I will be the pilot. I'll take care of you."

"How do you know if the mission is approved, you will be chosen to be the pilot?"

"I can tell that Reg, I mean Vice High Governor Jones, is sympathetic. If the mission is a 'Go' by the High Governor, then the Vice High Governor selects the flight crew. I am pretty sure I can persuade him to choose me. Reg wants me to prove my wings."

"You make it seem so simple."

"We need to present this proposal by next morning. You work with my Dad; I will work with Reg. Pick a world now, and then overnight, draw a list of mission mates. I want to give this to my mother as soon as possible."

"I don't know what to say."

"Save your best words for my father; I am already on your side. It's him you need to convince. Now let's go back inside and get to work. Drink some of Dad's beer (be sure to let him know how much you like it). Also, try the Rocky Mountain oysters; they taste like real testicles."

"Yes, ma'am.

Chapter 9.

The next morning, Yong felt a ringing in his head. It was Tessa accessing his implant.

"Wake up and get dressed. I am coming by in 10 E-min. to pick you up. We are meeting my mother at the *Peregrine Perch*."

Hastily washing and dressing, Yong left his room, after debating whether he should give Alison a message. Deciding not to, Yong took the lift to the lobby of the Falcon Hotel, and stepping outside; he found Tessa on an airscooter dressed in her Fleet uniform under a flight jacket. She had a helmet on and was holding another and a fleece jacket in her gloved hands.

"Put these on and get behind me. By the way, good morning."

Seated behind Tessa in the scooter, she bade Yong to wrap his arms tight around her waist as they took flight. As they soared overhead, in a dizzying flight between the mountains, Tessa shouted loud enough for Yong to hear.

"Mom has agreed to the mission. You did a good job on Dad. The world you chose, Bellerophon-b in the 51 Pegasi system, is perfect. The primary star is Sol-like (G2 on the Main Sequence), but because of the extensive zodiacal light surrounding the system, only the large Jupiter-like planet had previously been detected. Bellerophon-b is located beyond the edge of Human Space, but it is close enough for a mission to verify its suitability. What we need to convince them is that you and I will head the team that you select."

"Why am I essential?"

"Don't you want to receive full credit for your work? If you let someone else be the Chief Mission Scientist, it will be their name instead of yours associated with RIPE."

"What about you? Are you sure you will be the pilot?"

"I have been talking to Reg. He is considering it. This will be a great way to begin my Fleet career. We will make a great team. Aren't you feeling it?"

"Yes, I am," answered Yong, squeezing tightly around Tessa's waist.

Chapter 10.

After what seemed to Yong, a harrowing flight, Tessa took them down between the mountains to a picturesque valley that, in the center, stood what appeared to be a ski lodge. Alighting from the landed scooter, Tessa took off her helmet and signaled for Yong to do the same.

"My mother likes to ski in the morning before her work day starts. We are to meet her inside the *Peregrine* for breakfast. The lodge is a rec center for Federation personnel, especially for those who like to indulge in outdoor stuff. We are right on time. I think that's her coming down the mountain now."

As Yong shielded his eyes from the morning sun, he could discern shapes upward on the slope. As they came closer to the bottom, he could see that the lead skier was a tall woman in powder blue ski tights, a white sweater, and white ski boots. Flanking her were two other skiers, a woman, and a gentleman. Surrounding the trio was a troop of guards on skis – presumably the High Governor's Elite guardsmen. As the cohort came to a halt, Alyie Starstriker took off her helmet, letting her red hair fall freely in the brisk air. Yong gasped at how magnificent she looked, finding it difficult to judge who was the more beautiful, she or her daughter.

Sighting Tessa and Yong, Alyie waived them to join her party. She had skied down the mountain with her mother, Tanita, and Asher Weinbein.

"Good morning Tessa and Dr. Gwan, sorry to get you up this early, but my schedule is packed today."

"Morning, Mom, Grands, and Asher. You all looked super on the slope."

"Asher and I thought we'd take advantage of the opportunity while we are visiting," said Tanita. "I haven't skied since we lived on

Mars when Alyie was growing up. It is a lot easier here not needing oxygen tubes."

"Your form is impeccable, my dear," said a smiling Asher, who himself appeared very accomplished on skis.

"Good morning Madam High Governor, Commander Starbright and Mr. Weinbein," Yong stammered.

"Since you and Tessa are now apparently a team, I think in private conversations, you can refer to me as Alyie. May I call you Yong?"

"It would be my pleasure Madam High Governor."

Alyie then ushered all inside the lodge, where they proceeded to the High Governor's private dining room, where Jaynis and Rufus Steinhaus were waiting at a large table set for breakfast. Accessing her implant, Alyie messaged a group of autobots to serve breakfast consisting of flat cakes, slices of beef, blueberries, clotted cream, and juice and hot drinks.

After eating and making light conversation, Alyie used her implant to activate the holographic screen in the center of the room, where a suspended image of the grizzled but dignified head of Reg Jones was projected.

"Good day, Reg," Alyie beamed. "I hope all is well at your end of the Federation."

"Nothing out of the ordinary, I am happy to report," Reg responded.

"Now for the business at hand. After intensive lobbying by a certain individual and the authorization given by Vice High Governor Jones, I am ready to approve the Mission RIPEST (Remote Identification of Plausibility Environments System Test) to planet Bellerophon-b in 51 Pegasi system. Before I sign off, are there any concerns or questions?"

Rufus Steinhaus spoke, “Dr. Gwan has convinced us with his excellent presentation of the value to the Federation in using RIPE. But I wonder if our purposes would be as well served with a robotic mission to his candidate world rather than sending a human crew.”

“I have some rebuffs to that, Rufus,” said Jaynis, “but let me defer to Dr. Gwan.”

Yong was prepared for this objection to Tessa’s plan, and he responded, “A robotic spacecraft would be able to confirm some of the conditions as to why a candidate world received a high plausibility rating by RIPE, but it would not have the flexibility to test beyond the program. A human crew of specialists can make multiple tests of atmospheric chemical composition, sub-surface tectonic and hydrologic conditions, and temperature ranges. But most importantly, once RIPE’s plausibility of Bellerophon-b is confirmed, the team can initiate the Habitation sequence.”

“Yes”, added Jaynis. “I also support a human mission, especially for this proof of concept project. Once we gain confidence in RIPE, we can start Phase I with a full implementation of robot spacefreighters carrying all the tools we need for terraforming. Dr. Gwan, what are your crew and instrument recommendations?”

“I would like to serve as the Chief Scientist. Then we would need a Planetary Chemist, a Terra Engineer, a Hydrologist, and an Atmospheric Specialist. Of course, I would like Lieutenant Starbrook as Commander and Pilot. As for instrumentation, we would need a suite consisting of a magnetometer, mass spectrometer, seismometer, gravitometer, and radiometer. If we verify there is sufficient oxygen trapped in the topsoil, we will initiate its extraction by sweeping the planet with a molten salt electrolytic beam.”

Jaynis asked, “Dr. Gwan do you need an exobiologist?”

"No, Dr. Bottombrook, the data on gaseous emissions indicates there are no complex life forms on Bellerophon-b, at least none higher than insects."

"Reg," Alyie interjected, "have you chosen the flight officers?"

"I have assigned Captain Lucca Comani as the Commander, Lieutenant Starbrook as Co-commander, and Dr. Marisa Gomez as the medic."

"But Reg," Tessa protested, "I should be the Commander. This is my idea."

"I know, Tessa, but in consultation with your parents and respecting Fleet protocol, a first Lieutenant is never given sole command on her first mission, even if she is the instigator. We should not make an exception. You do understand, Tessa?"

"Yes, Sir, Vice High Governor."

"Now that we are all in accord, we can begin the planning," said Alyie. "Yong, your task here at HQ is to select, with the assistance of Jaynis, your fellow crew members. Once that is accomplished, and the mission is laid out by you and Tessa, you both will go to Teegarden to join Commander Comani for final preparations to launch Mission RIPEST. You have been granted indefinite leave from your University. I have informed Regional Governor Bernewithy that you will not be returning with her. I think she wants to speak in person to you before she gets on her flight to Paris."

"Yes, I should talk to her."

"I will arrange for her to meet you in the lobby of HQ."

Chapter 11.

As snow was falling outside the *Peregrine Perch*, Tessa turned to Yong just before they were to re-mount the airscooter to go to HQ. With her eyes gleaming and a wide smile, she placed her arms on his shoulders, exclaiming, "We did it. You were magnificent! Tomus was right about you."

"Right about me? I thought you said he didn't like me."

"Oh, that was about Alison. He did think you and I would be good together."

"Oh, he did? By the way, I am sorry you are not the Commander. The mission was your idea."

"It's okay. I had a feeling I wouldn't be the lead. Anyway, Commander Comani likes me, and I am sure he will let me do most of the flying. Besides, the important part is the planning which we can do together here on Earth. It will be great fun, don't you think?"

"Yes."

"Jump behind quickly. I see Mom and the others are leaving in her airbus. We'll follow in their wake. Hold on tight to me, and don't let go."

"Never."

Yong's meeting with Alison at HQ was terse. As Alyie, Jaynis, Tessa, and Rufus took the lift up, Yong waited in the lobby below. He was unsure of what he would say to Alison.

She appeared presently inside the aircar, which was evidently taking her to the spaceport. Stepping out of the car gingerly in her high-heeled boots, trying not to slip on the newly falling snow, Alison went inside the building. Holding her leather jacket tight around her shoulders, Alison greeted Yong with a waned smile.

“Well, just look how far you have come, my boy, in just a few weeks.”

“Alison, I am in a daze. I don’t know how I can thank you for all you’ve done for me.”

“Don’t mention it. Apparently, I have served my purpose. You are now in orbit in the Starbrook Universe. I hope you enjoy the ride.”

“I will keep in touch.”

“Yes, my boy, you do that.”

Chapter 12.

In the two E-w. that followed, Yong and Jaynis went through the process of selecting the scientific and engineering team for RIPEST. Yong knew some Planetary Chemists and Atmospheric Scientists from Chandrasekhar University and contacted them to see if they would be willing to participate in the mission. He received a yes from Dr. Anwar Mulafta, a Planetary Scientist in Yong's department. The Atmospheric people at Chandraeskhar University were all either involved in projects that couldn't be interrupted or had no appetite for space travel. Yong also sensed some professional jealousy in his meteoric rise to the top of the Federation's scientific hierarchy. Jaynis called on an old colleague, Dr. Millicent Dubwa, from his days at the Université de Paris, to fill the Atmospheric Scientist slot. Rufus Steinhaus volunteered to join the team as the Terra Engineer, and the final mission specialist selected was Dr. Nyra Fatima, a Hydrologist on the staff of the Scientific Institute.

Much of Yong's time was spent with Tessa planning the mission, fine-tuning the instrument specifications, ordering the selected instruments, and plotting the mission's trajectory. Since 51 Pegasi is over 50 l-yr. away, the trip would involve, at a minimum traversing through at least two hyperknots. Tessa did some optimization simulations and found what she concluded was the best course. She sent her recommendation to Captain Comani, who quickly approved the route.

A typical day found Tessa greeting Yong in the office that was assigned to him on the sixth floor of the Headquarters building bringing coffee, croissants, and fruit from the Federation Commissary. She would always come bouncing in, bright-eyed in her uniform of white short-sleeved button-down shirt, black tie and tight blue trousers (with black stripes), and black boots. Most of the time, she tied her hair in a ponytail with a colorful blue or green ribbon. She always smelled so fresh and clean that Yong was exhilarated by her scent.

Tessa would buoyantly greet him, "Top of the morning, my favorite Chief Mission Scientist. Let's see what the Gwan-Starbrook team can accomplish today."

He was reluctant letting Tessa know his feelings for her. She obviously enjoyed working side by side with him, but did this bonhomie extend any further than the mission? If he expressed himself to Tessa and she rebuffed him, would that ruin their working relationship? Or worse, if she told her mother of his ardor toward her would he incur the wrath of Alyie Starstriker, who could bring the power of the entire Federation down on him. No, he concluded, he would have to be just content with their professional relationship, although her repeated declaration of a Gwan-Starbrook partnership gave him hope.

Chapter 13.

When all their work was accomplished, Tessa and Yong boarded a spacecruiser along with Rufus Steinhaus and Nyra Fatima that would take them on a four E-d journey to Military and Security Headquarters on Planet Teegarden to rendezvous with Commander Comani and the remaining members of Mission RIPEST. Steinhaus was a tall, gray-haired older man with a white beard, while Fatima was a brown-skinned slender woman in her 30s with dark brown hair and eyes.

Yong had only a brief experience with hyperspace, so Tessa sat next to him, holding his hand during the jump, through and out.

"It is not so bad, is it?"

"It helps to have company," Yong replied, trying to suppress his nausea.

Spacecruiser *Seleucus* was nearly completely outfitted with the instrumentation and equipment when the party from Earth arrived on Teegarden. Tessa joined Lucca Comani to finish the preparations. Since this was not considered a military mission, the spacecraft was only outfitted with a laser cannon for defense, although all personnel were outfitted with a laser pistol and/or laser rifle. In the *Seleucus*' cargo bay were two space pods for surface excursions. Data showed Bellerophon-b to have a thin oxygen atmosphere, so breathing gear and helmets were essential for all personnel.

On launch day, the eight-person team was greeted by Vice High Governor Jones as they were about to board *Seleucus*. Reg shook Yong's hand and hugged Tessa long and tight.

"Have you said goodbye to your parents and thanked them for believing in you?"

"Yes, Reg. I have."

"You know they have a lot of faith in you."

"It is Dr. Gwan they should have faith in. This will be a historic mission because of his work."

"Well, then, I wish you both great success and safe travels."

"Thanks, Reg, I mean Vice High Governor," said Tessa while winking.

Chapter 14.

Lucca Comani and Tessa sat in the command deck as *Seleucus* lifted off from the spaceport. The Captain was a man of middle age with a robust built, light skin, brown colored hair, and eyes who Tessa had served under when she was finishing her commissioning. Marisa Gomez, the medical officer, had recorded baseline physical data from all the crew members prior to the flight and was using it to monitor any changes in their physiology throughout the mission. She sat alongside Yong in the passenger cabin. She was a 30 Earth-year-old petite dark-haired woman who seemed almost as nervous as Yong.

"Dr. Gwan, is this your first deep space trip?" asked Marisa.

"I have made the run from Earth/Mars to Teegarden a couple of times, but, yes, this is my first beyond the two Federation HQs."

"Well," chimed the physician, "you have company, for I am also participating in my first major mission. On the other hand, Lieutenant Starbrook seems an old hand, despite her young years."

"I think it is genetic with her," mused Yong. "She also was probably weaned while going through a hyperknot. What deep space experience do the other members of the mission have?"

"Mr. Steinhaus, Drs. Mulafta and Dubwa both have had prior exoplanet in-situ experience, but I think this is the first for Dr. Fatima. Of course, Commander Comani is a veteran of these types of excursions."

When meals were served by the spacecraft's autonomous food service, Tessa joined Yong, taking off her flight jacket, helmet, and gloves to sit opposite him. Looking at what served as dinner aboard *Seleucus*, Yong couldn't help but compare it to the haute cuisine he had on Earth with Alison. What a difference in quality and taste, but the company more than compensated for it.

The 50 l-yr. distance to the 51 Pegasi system took over three E-wk. Lucca and Tessa carefully guided *Seleucus* through two hypeknots without undue stress on the passengers. As the spacecraft emerged from the second jump from hyperspace, the dust cloud that was surrounding this system could be seen ahead. With force fields activated, *Seleucus* safely passed through the area whose zodiacal light was obscuring Bellerophon-b. The planet was located 1.6 A.U., safely in the designated habitable zone, executing a 1.5 E-yr. orbit around its star. As the day side of Bellerophon-b (its diurnal rate was nearly 1 E-d. with a 20° tilt) came into view, the Mercury-sized world showed a dark gray surface mottled with white spots. In another E-d., the team began RIPEST in earnest by initiating a deceleration burn to go into a high orbit around Bellerophon-b while taking readings with the onboard instruments.

Observing readings from the mass spectrometer, Dr. Anwar Mulafta, a cheerful gentleman in his 50s with salt and pepper hair, announced, "I am detecting trace metals and substantial oxygen in the regolith. Dr. Gwan, it looks like RIPE was correct about this planet's potential for producing a more sustainable environment. We can begin to start sweeping the planet with the molten salt electrolytic beam as soon as our survey is completed."

"I am seeing evidence of trace gases of oxygen and nitrogen in the atmosphere, but no H2O," noted Millicent Dubwa, a pensive dark-haired, middle-aged woman of medium height and built.

Using the remote sensing hydrometer, Nyra Fatima reported, "There is almost no surface water, but the data shows that there is sufficient groundwater about 10 cm. below the surface."

"Folks, I think it is time we go down for a closer look," announced Commander Comani as he and Tessa brought *Seleucus* to a lower orbit.

"Requesting permission to prepare the pods for surface landing, Commander," Tessa responded.

"Proceed, Lieutenant."

Chapter 15.

Since *Seleucus* was equipped with two surface pods, the plan was that Lucca and Tessa would each fly one carrying a trio of the mission specialists to survey a part of the planet. For safety reasons, both pods would not descend to the surface simultaneously. Alternately, Lucca or Tessa would drop off their passengers while the other stayed in orbit on *Seleucus*. When one of the pilots returned to the spaceship, the other would take the remaining trio to the surface. In this way, it was hoped that the verification process would be able to encompass a large part of the planet's environs in an expeditious time frame.

For the initial landing, Commander Comani, in Alpha Pod, took Rufus Steinhaus, Nyra Fatima, and Anwar Mulafta with him. Remaining on-board *Seleucus* were Tessa, Yong, Marisa Gomez, and Millicent Dubwa.

"Why didn't you volunteer to be on the Commander's pod? Don't you want to be the first to set foot on your world?" Tessa wondered.

"It is not that important," Yong replied, not mentioning his real priority was to ride on Tessa's pod.

Alpha Pod executed a gentle sweep above the surface before landing gently on what looked like a flat, barren plain. Once the specialists had disembarked and extracted their equipment from the lander, Commander Comani lifted off, making his return to the spacecraft.

Tessa waited until Lucca was back aboard *Seleucus* before telling her trio to prepare for their trip. Helping Yong and Marisa into their space suits and helmet, she gave them and Millicent a thumbs up when they were set to board Beta Pod.

Flying to the opposite hemisphere from where the first group landed, Tessa's party was dropped off in a shallow canyon. Walking easily in the 0.6 Earth gravity, Tessa skipped along, gazing skyward at the dim yellow primary star above the orange-red horizon.

"It looks pretty cool here, Yong, don't you think? Everything you predicted seems correct."

"I can't believe I am here, walking in this world."

"After it is developed, maybe we can make a return visit, a sort of mini-vacation for both of us."

"I would like that."

As the science team began the soil sampling (which Marisa agreed to help with) and the seismologic readings, Tessa wished them good fortune and returned to *Seleucus* to join Lucca in monitoring the data the mission specialists would be transmitting.

Chapter 16.

As data streamed into the systems aboard *Seleucus*, there were indications of microbial life in the soil. Also, the hollowness of the surface terrain suggested there were extensive underground caverns. Rufus Steinhaus sent a message they had found an entrance to one.

"From our sonic sensor readings, it appears to be a large cave with multiple passageways. We are going in to investigate."

"Copy that, Alpha Team Leader. We will track you with the infrared scope," answered Commander Comani.

Yong also confirmed in a message to Tessa that the soil has primitive life forms, but the hydrologic data suggests an intricate underground network of channels.

"Tessa, it is as if these subterranean waterways were designed rather than just formed by geological processes."

"Yong, if they are constructed, the proper name for them are canals instead of channels. This may be evidence of some sentient life forms either in the past or still present."

"How can that be, Tessa? All the biochemical evidence suggests no life form beyond insects."

"But, don't colonies of ants build elaborate structures. What if evolution on this planet favored the genus Formicidae or mutations of such? Wouldn't that explain the canals or the passageways that Alpha Team is reporting?

"Yes, but where are they, Tessa?"

"They may not be active at this time of day. Yong, I have a bad feeling about this. I think I better come and pick you up before you stumble into something dangerous."

"I am surprised at you, Tessa. I thought you were fearless."

“The prime lesson we learned at Fleet Academy was to be afraid of anything you don’t understand. Gather the others; I am coming to get you. I will inform Commander Comani.”

Just as Tessa was about to request Lucca’s permission to make a landing, she noticed how pensive he looked while monitoring the infrared signatures of Alpha Team.

“Something is amiss, Lieutenant. I am reading that my three scientists are moving frantically below the surface as if they are in some kind of danger. I don’t know what to make of it.”

“Commander, they may be under attack from the resident life forms on this planet.”

“But I am not getting any other heat signatures around them.”

“That could be because they are members of an ant species who don’t emit thermal radiation.”

“That is incredible. How can that be?”

“My team has noticed intricate waterways that don’t seem natural. We may be dealing with an organized civilization that would not see our presence on their planet as benign.”

I don’t know what to think, but to be on the safe side, I am going down to retrieve my team.”

“I also request permission to leave and pick up Beta Team, Sir.”

“Denied, Lieutenant. You need to remain aboard *Seleucus* in the command deck until I return.”

Chapter 17.

Tessa waited anxiously as Comani took Alpha Pod back to the location of his science team. She feared a delay in her getting Beta Team back was increasing their chances of coming under an assault. Suddenly as she was debating whether to disobey the Commander's order, he sent a distress signal to her implant. Taking the spacecraft out of orbit, Tessa brought *Seleucus* right above the location where Alpha Team had been deposited. Breaking into the spacecraft's com system, Lucca cried, "We are under attack from these monsters. They have torn apart Steinhaus and Mulafta. I am carrying out Fatima."

Looking through the ship's visual telescope, Tessa could see Commander Comani dragging a wounded Nyra Fatima as he was besieged by only what could be described as human-sized red ants. They reminded her of the harvester ants that infested her parents' home in Colorado Springs. Lucca was using his laser pistol to fend them off as he tried to make his way to Alpha Pod. The ants surrounded him and Nyra and repeatedly stung the two humans in the backs and chests.

Tessa left the command deck to don her helmet and enter Beta Pod to attempt a rescue. She stopped when she heard Lucca say, "Forget us; save the others."

Rushing back to the telescope, she witnessed the horrid scene in which the swollen bodies of Comani and Fatima were lifted up by two giant ants who decapitated them with their mandibles.

Trying to compose herself and resisting the urge to vomit, Tessa realized her sole mission now was to save Yong and the others. She jumped back in the captain's chair and sent *Seleucus* on course to the lower hemisphere of the small planet.

Hailing Yong, Tessa cried, "Prepare for extraction. I am sorry I am late. There has been a hostile attack on Alpha Team."

"Tessa, you were right about life forms here. Along the rim of the canyon, there are thousands, maybe more, of giant-like ant creatures surrounding us. I think they are preparing to attack. Hurry, please."

Realizing the situation was dire and she might not be able to land and take the team back up in time, Tessa had a brief panic attack. Controlling herself, she wondered what her mother would do in this situation. The answer came immediately.

"Yong, I need your advice," Tessa announced. "We have some weapons on the spacecraft, but nothing that can stop an advancing army of ants. (I wished we had a herd of Mom's Beelzebub, jumbo-sized.) Do you have any suggestions?"

"Tessa, I am not good at snap decisions, especially now."

"Think, Yong. Use that great brain of yours. Is there anything we can do to prevent them from coming down the canyon, or at least holding them off until I rescue you?"

After what seemed an enormous wait, Yong burst forth, saying, "The electrolytic salt beam may do it. Once heated, it will cause the soil to be molten. Spray it around the rim of the canyon. It may just work, at least for a little while."

"Roger that, partner."

Tessa then sent the spacecraft in a looping pattern around the canyon and initiated the process which sprayed the surface with the salt solution. Then using the laser cannon, she heated the saturated soil until a ring of fire was produced. Tessa could see that the members of the ant army were either caught in the flames or retreated back away from the rim.

Realizing this was only a reprieve from their assault, Tessa put *Seleucus* in autopilot and attaching her breathing tubes in her nostrils, she donned her helmet and gloves, taking care to make sure her laser rifle and pistol were fully charged. Getting into Beta Pod, she jettisoned it from the spacecraft and made a soft landing in front of where she had deposited the team earlier.

The three scientists rushed to the pod as Tessa jumped out to assist them. Suddenly, ants sprang from below where there must have been underground tunnels. One attacked Millicent Dubwa, seizing her right arm in its mandible. Tessa discharged her rifle and severed its head off. As Marisa and Yong went to help Millicent, another ant came charging forward, grabbing Marisa by her left leg. Tessa aimed at the creature's body and set it aflame. Millicent's arm was dangling loosely and Marisa was bleeding profusely in her lower leg as Yong and Tessa helped them both into the pod, as more ants rushed toward them.

"Get in. I'll hold them off until it is safe to take off."

"No, Tessa. You get in first. You're the pilot. Just tell me how to use your rifle."

"OK. Just point and squeeze. It will do the rest."

As Tessa hopped into the pod, Yong fired at a creature about to climb on top, setting its head on fire. He sprayed the laser rifle in a sweeping direction to ward off any further immediate attacks, and responding to Tessa's insistence; he climbed aboard.

The soil must have cooled as most of the rest of the army was now advancing down the canyon. Fortunately, the pod lifted off before the attackers reached them, and they headed back to the spacecraft with two wounded passengers.

Chapter 18.

Millicent needed immediate attention as she was in danger of losing her arm. Tessa and Yong put her on the spacecraft's operating table, and the autodoc system, once activated by Tessa, diagnosed her condition and went to work sealing the wound and sowing her arm back together. Tessa applied a tourniquet to Marisa's leg and injected her with antibiotic solution to ward off infection. Once the wounded were stable and resting, Tessa contacted Fleet Command to inform them of the calamity that befell the mission.

As she was speaking to the duty officer, Reg Jones broke in, "Tessa, are you alright?"

"Yes, Reg, Yong, and I are bruised but not injured. I think both Drs. Gomez and Dubwa will pull through. We have; I am sorry, we lost the others. Should I attempt a recovery mission?"

"Negative, Lieutenant; just keep your wounded alive and come back safe."

"Yes, Sir."

On the return trip, both Millicent and Marisa gradually recovered. Dr. Dubwa's arm was saved, and, in time, she would regain almost its full motion. Dr. Gomez hobbled around the ship on a crutch but was able to take over the duties of administering to Millicent.

Yong spent a good deal of time by himself, brooding over the terrible turn of events. At first, Tessa let him be alone, but seeing his despair, she decided to leave the captain's chair to join him in the cabin. Sitting beside him, she put her arm around his shoulder.

"Don't beat yourself up over this. How would your algorithm anticipate such a bizarre genetic turn?"

“That’s where RIPE failed. It used studies that scanned for LAWKI, namely ‘Life As We Know It’. These programs didn’t anticipate the evolution of Formicidae. I am a complete failure.”

“Don’t talk like that. Yes, your algorithm was not accurate, but that doesn’t mean it can’t be improved. My Dad says science is always a process of guessing about nature. Humans propose, learn from mistakes, and get better. You will as well.”

“I don’t know if I will get a second chance.”

“Listen, if anyone is responsible for what happened, it’s me. I am the one who pushed for this mission.”

“Tessa, if it weren’t for you, Marisa, Millicent, and I would have been fodder for those creatures. We owe you our lives.”

“We both made a difference. The Gwan-Starbrook partnership is alive and well. Whatever happens, when we return, we will face it together,” said Tessa as she clenched Yong’s right hand in hers and pressed her cheek against his.

Epilogue

At the service for the fallen of the RIPEST Mission in the Memorial Gardens on Teegarden, the whole Starbrook family (Alyie Starstriker, Jaynis Bottombrook, Tessa and Tomus Starbrook, Tanita Starbright, Jaynis's father, Chadwick Bottombrook and sister, Bernice) joined Vice High Governor Jones in tribute to Lucca Comani, Rufus Steinhaus, Nyra Fatima and Anwar Mulafta.

High Governor Alyie Starstriker, in full uniform with medals, sash, and sword, spoke these words as a eulogy,

"In any new endeavor, there are risks that need to be taken if we as a species will advance forward. It is the bravest of us who assume this burden for humankind, knowing full well the costs could be their very lives. But it is through this bitter sacrifice we gain the knowledge and wisdom that enables humanity to survive and thrive in this often hostile Universe. To Lucca, Rufus, Nyra, and Anwar, we promise that their courage was not in vain, and we dedicate our future explorations in their memory."

After the ceremony was over, Tessa invited Yong, who was in attendance with Marisa Gomez and Millicent Dubwa, to join her family for luncheon in the Banquet Hall at Teegarden Headquarters. The family and friends of the fallen were to be honored at the event. Yong had never seen Tessa so formally dressed in a white double-breasted jacket, white gloves, trousers, and a cap. Only her tall black boots broke the color scheme. Seated at the head of the table were Alyie and Reg, titular heads of the Federation. Tessa had Yong sit between herself and Tomus, who now was in his natural skin color and his hair cut short, was eager to engage in conversation.

"From my mother's speech, it looks like you two are fully absolved of what happened on Bellerophon-b. Mom has a gift of making a positive out of the most dreadful circumstances."

"Yes, I get to keep my commission, and Yong can go back to Mars, although I think Dad is interested in keeping him at the Scientific Institute on Earth. He knows how brilliant he is."

"My sister tells me you are also quite handy with a laser rifle when need be."

"I only used it out of desperation. If it weren't for Tessa, I wouldn't have known how to work it."

"Dr. Gwan, I now see I may have made a misjudgment. There appears more to you that meets the eye. But I was correct about one thing."

"What is that Mr. Starbrook?"

"You and Tessa make a great team."

"We feel that way," Tessa said as she leaned over to plant a soft kiss on Yong's lips.

A flushed-faced Yong could only nod in full agreement.

The Images on the Wall – Part I.

To hold as 'twere the mirror up to nature: to show virtue her feature, scorn her own image, and the very age and body of the time his form and pressure.

Hamlet, Act III, Sec. 2. W. Shakespeare

Prologue

As the submersible vehicle glided beneath the bright blue water, moving images appeared in the transparent windows onboard. Aquatic life forms of different colors, sizes, and shapes surrounded the two travelers displaying the fecundity and diversity of this undersea domain. As the submersible plunged deeper and moved closer to the shoreline, the sonic transmitters on the craft recorded a structure straddling the ocean floor. Upon approach, it was seen to be the opening to an underwater cave. Daring to enter the mouth, the travelers set the submersible down before the entrance and went outside in their cave diving gear with oxygen tubes and canisters. As they swam further inside the cave, the water became shallow, and they were able to stand upright. Counteracting the engulfing darkness with illuminated handheld lamps, they followed a tunnel-like path to an expansive room-like enclosure. As the male member of the duo shined his beacon on a wall, there appeared a set of drawings left by previous occupants. Getting the attention of his female companion, the two of them crept closer to examine the artifacts and made a revelation that would change the history of humankind.

Chapter 1.

It was a matchless day, with the Kanvaian sky and the Andaman Ocean in competition for the bluest. Tanita Starbright and Asher Weinbein returned to the surface after their excursion below, aligning their submersible alongside the *Penelope*. Tanita, now in her late 70s, remained a striking-looking woman with brilliant white hair, a lean agile body, and startling blue eyes. Her companion, Asher, was the epitome of elder masculinity, sporting now a trim silver beard and mustache that complemented his bronze-toned muscular frame.

They were on Kanva visiting Tanita's daughter Alyie Starstriker, the former High Governor of the Sidereal Federation, and her eternal partner, Dr. Jaynis Bottombrook. Since Alyie's term of office had ended 2 E-yr. ago, she and Jaynis departed Earth and accepted her friend Queen Revita Jadeite's offer to settle in Kanva as a temporary residence while they decided what to do next in their lives. Revi had made available for them a chateau on the outskirts of the capital city Inmar. Tanita, Professor Emeritus at the Federation Fleet Academy, and Asher, who had retired from his position as Hotelier at the Hallows Inn on Cognito, made this trip a vacation and were excited to take advantage of the water activities available over Kanva's vast ocean.

As she and Asher were quickly getting out of their thermal wet suits and removing their flippers, and dressing in their on-board clothes, Tanita realized what she and Asher observed below needed immediate attention. However, the news required discretion because of its possible incendiary nature. Asher programmed the *Penelope* for a course that would bring them to the mouth of the bay where the tiny island, Ismar, was located. Tanita knew that today Alyie was out horseback riding with Revita.

Docking *Penelope*, Asher, and Tanita walked the short distance to the Royal Stables, where they were told by the stable staff that Queen

Revita and Madam High Governor Starstriker should be returning soon. The couple sat on a mock wooden table outside and were served scones, strawberries, clotted cream, and tea. The two friends returned presently in high spirits, flushed from their time together. Revi was on a milk-white mare, Phaedra, while Alyie rode Shera III, a dark-hued stallion, a descendent of her favorite mount. The two, now middle-aged women, made a brilliant pair. Revita's finely chiseled face was smooth and delicate as ever. Her lustrous black hair peeked under her helmet, and she wore a white silk kerchief tied around her neck. Alyie had lines on her brow and crow's feet beside her eyes to show for the 15 E-yr. she served as High Governor. She was also a little wider around the hips but still a spectacular-looking woman. Her now flaming red hair was tucked under her helmet, and she wore a green ascot covering the nape of her neck. The two women were clad identically in cotton white short-sleeved tee shirts, bright white breeches, and tall black riding boots with spurs.

Seeing her mother and Asher on the island, Alyie immediately sensed something was afoot. Dismounting from Shera III, she went up to where they were sitting and kissed Tanita.

"Mother and Asher, how delightful to see you. It's great to have you both joining the Queen and I for breakfast, but weren't you going to spend the day on the ocean?"

"Hello, my dear," Tanita replied.

Asher took Alyie's hand and kissed it, saying, "Bonjour, Madam High Governor."

Revita had handed the reins of the two mounts to the stable grooms and joined the threesome.

"Your Majesty," bowed Asher while Tanita gave a short curtsy.

As the two friends joined the couple at the table, staff appeared, serving the Queen and the former High Governor.

"Our visit here, although enchanting, is not social," Tanita began. "Asher, please."

"Your Majesty and Madam High Governor, when Tanita and I took the submersible below the surface, we found an underwater cave near the shoreline. Being curious, we left the vehicle and set out to explore the cave."

"Mr. Weinbein and Tanita, I would say you are not only curious but fearless to venture into such an unknown area. I applaud your courage," remarked Revita.

"After traveling through what was a passageway, we found ourselves in a large room that, once illuminated, displayed what I can only say were drawings on the walls."

"After viewing them," Tanita interjected, "I felt it was vital that Alyie see them. Asher will show you the images we recorded."

"Mother, why didn't you just send me the pictures you took of my implant and continue on the water?"

"Madam High Governor, due to the sensitive nature of their content, your mother and I thought it best for security reasons that we show them in person."

Turning to give a sly smile to Revita and gently touching her friend's left boot, Alyie chided them, saying, "Let's see these not-for-public viewing images. Revi and I are, if nothing else, always discreet."

What was shown was a series of what could be called murals in which crudely drawn human-like figures were involved in activities such as hunting, fighting, and harvesting. There also appeared some worship of a deity that looked like Kanva's star, Trappist-1.

Alyie's frivolity ceased immediately, and she turned to Revita, asking, "Do you have any knowledge of who produced these drawings?"

“None. As you know, Kanva was settled by humans two E-c. ago. Since its benign primary star and the four E-d. synchronous rotation produces a balmy climate with a robust greenhouse effect and an oxygenated atmosphere. The planet needed little terraforming to become habitable. There were fauna, and flora found when it was first settled, but there was no evidence of higher life forms.”

“Well, these pictures,” Alyie concluded, “would indicate there is evidence of not only higher life forms but hominids.”

“You know what this may mean?” Tanita said after taking a sip of her tea.

“Earth may not be the cradle of humanity,” Revi responded.

“We thought,” Asher interjected, “it would be best to bring it to you, Madam High Governor, before it becomes public knowledge. We fear the upheaval in the Federation this news will bring.”

“Oh, Alyie,” Revita gasped, “this may cause undue all the achievements you accomplished in making Earth regain its importance in the Federation.”

“The problem, Mother and friends, is that I am no longer the High Governor. I am now a civilian and a guest on Kanva of Queen Revita. I have no Federation authority.”

“You may be out of the office, my dear daughter, but there is no one in the entire Human Space who is more adept in dealing with a situation such as this. Both Asher and myself don’t think the current person who occupies the High Governorship is up to this task.”

“Okay, let’s first try to ascertain the authenticity of these drawings. If they are shown to be contemporary, then we need not pursue it any further. I know someone who can be counted on with absolute discretion who is outstanding in antiquities. Let me consult with Jaynis first. Mother, I may ask you to be the go-between.”

Chapter 2.

After the meeting with Tanita and Asher, Revita flew Alyie back to the latter's chateau. Disembarking from the airjet on the foreground of her residence, Alyie went right inside the front door, dispensing with a trip to the mudroom to get out of her riding clothes. Clapping loudly on the marble floor with her black boots and silver spurs, her riding crop slapping against her right boot, Alyie was greeted by her senior attendant, Greta Gudwig.

"Good day, Madam High Governor. I hope you had an enjoyable ride."

"It was brilliant, Greta. Is Dr. Bottombrook available?"

"I believe, Madam Governor, he is still in bed sleeping. Would you like me to direct an autobot to wake him?"

"No, Greta. I will reserve the pleasure of rousing him all to myself," Alyie coyly responded.

Tucking her riding crop snuggly inside her breeches such that its keeper rubbed against her groin, Alyie galloped up the spiral staircase. She softly opened the door to the premier bedroom, finding Jaynis indeed still asleep, his light brown and gray hair tousled over his face. Quietly as possible, Alyie unfastened her ascot, yanked off her boots, extricated the riding crop, glided out of her breeches, removed her riding hat, and let her hair down. Climbing into bed, she snuggled next to Jaynis and gently stroked him. Feeling her presence, he opened his eyes, smiled at the stimulation, and kissed her firmly on the lips. They embraced and tumbled in bed, with Alyie getting the upper hand.

"It seems horseback riding with Revita brings out the beast in you. Am I now your next mount?"

"Think of me as a gentle beast, and please be a dutiful mount to your rider," Alyie laughed.

"Yes, mistress," Jaynis replied as he came to full attention.

When their bed business was concluded, Alyie, nestling her head under Jaynis' arm, explained that her wake-up call was more than just conjugal.

"I think there is a problem. Tanita and Asher discovered an underwater cave with primitive drawings that may indicate a previous hominid civilization on Kanva."

"Wow! You call this a problem," said Jaynis as he nibbled Alyie's right ear. "I say this is a great discovery."

"Yes, but I fear it will open up divisions in the Federation as to Earth's preeminence. I want this find to be investigated thoroughly before it is publicized. And when so, the announcement must be handled with the utmost sensitivity. "

"I hate to remind you, my love," Jaynis replied while fondling Alyie's breast, "but you are no longer the High Governor."

"I am aware of that, but my mother believes the present occupant of that office is not capable of dealing with such a volatile issue."

"True, but how will you keep it secret from the Federation hierarchy while requesting an investigation?"

"I have in mind someone we know who is interested in this kind of find who we can trust absolutely."

"I think I know who you mean, but is he ready for this type of responsibility?"

"I believe he is, but the only true test is to do it."

"Then let's ask him."

"I don't want the request to come from me. I will ask Tanita to serve as the conduit."

"Good. She will like that. Now, where were we on your riding lesson?"

"Now look who has become the beastie," Alyie laughed prior to planting Jaynis with a wet kiss.

Chapter 3.

Dr. Tomus Starbrook was at his office at the Ministry of Arts and Culture in Paris when he received a message from his grandmother on his implant.

"Hello, Tomus. How are you? Asher and I are having a great time on Kanva. Your parents send their love. We have uncovered some interesting artifacts that we would like you to look at. For reasons that will become clear, we would like you to come here in person to examine them. Your mother highly recommends you do this."

Unsure of the underlying meaning of the message, Tomus nonetheless noted the not-so-subtle order from his mother. This was unusual in that she normally lets him decide for himself his assignments. Now 23 E-yr. old, Tomus had left the Louvre Museum to take a position as an Assistant Director of the Ministry of Arts and Culture. He had shaken off his taste for exotic looks, with his now short-cut hair resuming its natural blond color. His skin tones were neutral, and he dressed in dark sports jackets, shirts, and slacks.

Contacting his one-up, the Deputy Director of Arts and Culture, Mr. Hashin Samak, Tomus asked for permission to travel to Kanva.

"Dr. Starbrook, what is your justification for leaving the Ministry? You know we have budgets to submit to the Federation, and there is our five-year audit report to prepare."

Mr. Samak was a 20 E-yr. veteran of the Ministry who had served under Alison Bernewithy when she was the Director. Tomus sensed a bitterness in Hashin in that he never rose to be the head. Tomus' close relationship with Alison and his own rapid rise up the ranks of the Ministry added to Mr. Samak's dislike for Tomus.

"I know this seems sudden, but much of my family is on Kanva, and they are planning a reunion. If I am not granted official leave, I will take my vacation time for this trip. Be assured I will continue to work on my assignments remotely."

"Then that is what it will have to be Dr. Starbrook. Please conclude your business quickly on Kanva and return as soon as possible."

"Yes, sir," Tomus responded, musing how things have changed since his mother was no longer High Governor.

Chapter 4.

In order to keep up the pretense of a family reunion on Kanva, Tomus contacted his twin sister Tessa to see if she would accompany him. Tessa was First Lieutenant on the *Falcon* Fighter Squadron based in New Cedonia.

Reaching her by video hyperlink while she was piloting her fighter, *Accipiter*, Tessa flashed a wide smile that shone through her flight helmet, saying, "Hello, Brother. It's great to see you. How are you doing ascending the Federation hierarchy?"

"Hello, back to you, Tessa. Being an Assistant Director of the Arts and Culture Ministry is not considered high on the ladder of the Federation."

"Yes, but you now are in the chain of command. You can pretend, but I know you and the Regional Governor have set things in motion for your career. She has big plans for you."

"Yes, Alison advises me on my future. But right now, I want to stay in the Arts world."

"Listen, don't misunderstand me. I want you to succeed. Since Mom and Reg have left, the Federation is slumping. I don't think High Governor Thakur is competent."

"Sister, shouldn't you be careful of what you say aboard a Federation fighter communicating on a Space Command hyperlink channel?"

"Oh, I have broken enough rules already in my three E-yr. military career; I don't think dissing the High Governor will matter."

"Before you get censured, let me tell you the purpose of this call. I have been requested by Grands and Mom to visit them on Kanva and do some art investigation. Would you like to join me? I am planning to go in about one E-wk."

“I would love to get together with you and the folks, but two big obstacles stand in the way. First, my patrol is on an almost two-week run. When I get back to base, I have some leave time, but I promised Yong we would spend it together. We haven’t seen each other in person in almost nine E-m. He is even traveling to New Cedonia to be with me. We have a lot of catching up to do if you follow my meaning.”

“I get it, Tessa. I may be inexperienced in these matters, but I am not naïve. Give my best to Yong. But if you change your mind, both you and Dr. Gwan are welcome. There may be some interesting things happening on Kanva; that, at this point, I don’t fully understand.”

“Oh, Tomus. You are the clever twin. Now, you piqued my curiosity. I will talk to Yong and see if we can join you later.”

“Great.”

Chapter 5.

Tomus was able to gather from a subsequent secure communication with Tanita that she and Asher had discovered some paintings on cave walls underneath Kanva's shoreline. Not himself an expert on such art, he contacted a former professor of his at the Université de Paris, Dr. Renard Fouya, a known hieroglyphologist, to accompany him on this investigation. Dr. Fouya, a boisterous middle-aged man with sparse curly hair and a rotund build, leaped at the chance to participate in this expedition.

"Professor Fouya, I have to ask for your confidentiality about our trip until we are convinced of the provenance of what we find on Kanva."

"Be reassured, Tomus, my dear boy, that I know enough about scholarly priority to keep a tight lid on the reason for our journey to beautiful Kanva."

As scheduled, in two E-w. Tomus and Renard boarded a private spacecruiser for their flight to Kanva. It was Alison Bernewithy, who, as Regional Governor of Sol System, was able to make the travel arrangements.

Alison, on a hyperlink video, wished Tomus a safe trip on the eve of his departure.

"Tommy, give my best to your parents, and remember, be careful of what you do and say publicly. The eyes of the Federation are on you."

"Alison, the only eyes I know are on me are yours. But that happens to be the only pair I really care about. Thanks."

The three E-wk. flight was uneventful, other than Dr. Fouya giving Tomus a seemingly endless rendition of the professor's curriculum vitae.

Arriving at the spaceport on Kanva, Tomus, and Dr. Fouya were met by a young woman who was standing in front of an airjet.

"Greetings, I am Melissa Turnberry, a member of the Queen's palace staff at your service."

Melissa, an attractive well turned out young woman (about Tomus' age), was dressed in the green colors of the House of Jadeite in a one-piece jumper suit with her green tights tucked into black boots. She had a black silk scarf around her neck that was secured by a jade pin. Her jet black hair, alabaster pigmented skin, and sparkling green eyes gave her a most exquisite look.

"Hop in," Melissa declared as she took their travel bags into the airjet. Once all three were on board, Melissa took the craft aloft, and in less than 30 E-min, they landed on the grounds of the palace.

As Melissa directed Tomus and Renard up the palace steps, she declared, "The Queen and your family Dr. Starbrook are waiting for you in the throne room. I will escort you, and then I will arrange for your belongings to be brought to your respective rooms. During your stay in Kanva, I will be your contact person. Let me know through the link I am sending to your implants if you have any questions or need my services in any way."

"Any services?" inquired Dr. Fouya.

"Yes, Dr. Fouya. Queen Revita wants you to have a most successful and satisfying stay on Kanva."

"This is going to be even better than I imagined," Renard whispered to Tomus.

Chapter 6.

The main hall leading to the throne room found Revita, Alyie, Jaynis, Tanita, and Asher waiting to receive the travelers. Rebina Ectot, the senior Queen's Lady-in-waiting, was also present, standing discreetly behind the Queen. Alyie came forward and hugged her son warmly, saying, "God, I have missed you. What happened to my little boy? You are now a man."

"It happens, Mother, but you have not aged at all. Living on Kanva becomes you."

"I see my son has learned flattery as part of his maturation. But yes, Queen Revita is a wonderful host; your father and I are thriving here.

Next, Jaynis came forward to give Tomus a firm handshake.

"My, you have grown since we last saw you. I now believe you are taller than your mother. Dr. Fouya, how wonderful it is to meet you. I have read some of your papers and marvel at your erudition and your ability to add complexity to the simplest topics."

"Greetings, Dr. Bottombrook. If you have time, I would be happy to present to you my latest discoveries of ancient human writings."

"A pleasure, no doubt, Dr. Fouya, I may have to forego. My mother-in-law and Mr. Weinbein have a tight schedule of expeditions they have planned for you and my son on the ocean."

The travelers then approached the Queen, who accepted their bows.

"Yes, there is much work ahead for both of you. But that can all be discussed tomorrow. You have traveled far, and I want you both to relax and enjoy the amenities of my country. I believe you have met Ms. Turnberry, one of my Ladies-in-waiting. You will find her delightful and efficient. I have tasked her to make your stay as wonderful as possible. Go to your rooms and rest. We will have a reception and welcoming dinner in the Banquet Hall when the shades turn the interior dark. Dress as you wish. We are all family and friends here."

Tomus and Renard were then escorted to their respective rooms. Tomus found his spacious, well-lighted, and attractive. Once he had unpacked and settled in, Melissa signaled for permission to enter to see if his needs were met and whether she could provide anything extra.

Tomus allowed her entry and said, "No, everything is perfect. Do you think I should dress formally for this dinner?"

"I know the Queen and the High Governor like to dress up for these occasions. If you excuse my impertinence, they are like school girls in friendly competition with each other. It's really so much fun to see them together. Your father and Mr. Weinbein will probably be understated in their attire. If you have a casual suit or sport coat and slacks, that should be sufficient for tonight. If you don't, our robotic tailor can make you one on the 3D fabricator in time for tonight's event."

"No, that will not be necessary; I have brought an appropriate outfit. Thanks for your advice. Good day."

"Before I leave, I have an issue I want to discuss with you. When I paid a visit to your colleague, Dr. Fouya, to see if everything was well with his accommodations, he became most forward with me. I can report his behavior to the Queen, but I fear this will taint your whole visit to Kanva. Can you give me suggestions on what I should be doing to discourage him?"

"Oh my, I had a feeling Dr. Fouya would act this way when he first saw you. I have already given it some thought. Let me handle this. It may involve a little deception on our part. If you are willing to play along with me, I think we can make him see that his efforts are futile because you have an interest in someone else."

"I see. You do realize that I am committed to the Queen."

"Yes, I surmised that. If our ruse doesn't work, then feel free to report his bad behavior."

"Thank you, Dr. Starbrook."

Chapter 7.

The dinner was what one would expect from a royal house. An elaborate setting with several plates, a dizzying confusion of silverware to use, and a profusion of crystal glass goblets. True to form, Revita and Alyie were clad in dresses, the Queen in an emerald full-length gown secured on one shoulder by a dazzling diamond broach, whereas Alyie's was snow white with a tight bodice to display the High Governor's cleavage. Both women were in high heels, the Queen's black and Alyie's white. Tanita appeared in a dark blue pants suit, black heels, and a white satin blouse draped over by a string of black pearls. Jaynis and Asher, as predicted by Melissa, were more informally attired, with the former in a buttoned-down blue shirt and black trousers and the latter in a brown sports coat over an off-white turtleneck pullover shirt and light brown pants. Also joining Tomus and Renard at the Queen's table were Rebina Ectot and Melissa Turnberry, with the latter seated between the two gentlemen from Earth.

At the appropriate time, Queen Revita rose and gently stopped the conversation by raising a glass of champagne in her elbow-length green gloved right hand, offering this toast,

"May our visitors find success in their investigation in searching for the truth while preserving the unity of the Federation. À votre santé."

Cheers were seconded by all present.

During the dinner, Renard kept trying to convince Melissa that there is something in his room that needed looking after. When she didn't acquiesce to his invitation, he reminded her that she was in his service on Kanva. Tomus could see she was getting uncomfortable with his persistent double entendre. Feeling he needed to step in, he turned to her, saying, "Melissa, you promised to give me a tour of the palace. Can we do this when the banquet is concluded?"

Picking up Tomus' cue, Melissa responded, "Of course, Dr. Starbrook, it would give me great pleasure to show you the beauty and grandeur of the palace both inside and outside. Mind you; it will take some time."

"Capital," said Tomus. "Dr. Fouya, you are welcome to join us, but I warn you we are scheduled for an early start tomorrow."

"No, my boy," said a subdued Renard. "I will just return to my room now."

Chapter 8.

When the automatic shades proclaimed the next day on Kanva, Tomus and Renard were summoned to breakfast with the Queen, Alyie, Jaynis, Tanita, and Asher in the Emerald Room. A sumptuous spread of eggs, flat cakes, and a cornucopia of fresh fruits with cream, freshly prepared by the Royal Kitchen, were laid out on the sideboard, accompanied by hot beverages.

After a suitable time, Asher began the discussion providing Tomus and Renard with details of the underwater discovery made by Tanita and himself.

Looking at the recorded images, Renard Fouya commented, “They look authentic, but to convince myself, I will need a closer look, of course.”

“That is what we are prepared to give you, Dr. Fouya. I know you and Tomus are not experienced divers, but we have fabricated an underwater suit that will adhere to your body and allow you both to leave the submersible and come as close as you wish to the walls of the cave where the murals are displayed.”

“If the suits are ready, I suggest we get started immediately,” Tomus pronounced.

“Good; after you refresh yourself, Tanita and I will drive you to the water’s edge, where everything we need is waiting.”

Tomus and Renard were wished good luck by Queen Revita. Jaynis and Alyie shook Renard’s hand, and each, in turn, hugged their son.

Alyie held Tomus proudly before her, saying, “This may be an important mission. I have absolute faith in you.”

“Don’t worry, Mother. I won’t let you down.”

"Or the Federation, my son."

Escorted by Melissa Turnberry to a lavatory, Tomus and Renard were brought clothes more suitable for boating. In the changing room, Renard asked, "I am jealous of you, my boy. I thought I had the inside track on that delightful Ms. Turnberry, but it appears you stepped in front of me."

"Dr. Fouya, we are here on an investigative mission. You and I have no time for dallying with the Queen's staff. It would be unprofessional."

"You only say that after you had a night of delights. I am on to you, my boy."

"Let's just do our job, Professor."

Chapter 9.

Asher and Tanita guided *Penelope* to the location near the shoreline where they had discovered the cave. Fitting Tomus and Renard in their transparent oxygen-filtered body suits, Tanita steered the submersible while Asher stayed on board the *Penelope.* As the submersible dived below, its three inhabitants marveled at the underwater life contained in the Andaman Ocean. Reaching the bottom, Tanita had the submersible slowly approach the cave using sonar to navigate. When the craft was at the mouth, she signaled Tomus and Renard to prepare to disembark. The three moved and swam carefully into the darkness, with Tanita shining her beacon before them. Finding the passageway that led to the underwater cavern, Tanita stood up in the shallow pool with Tomus and Renard right behind.

Tomus found he had to duck his head to enter the hall, but once inside, there was a high ceiling that allowed him ample room to stand upright. With Tanita sweeping her lantern across the walls, Tomus and Dr. Fouya could view the artwork. Getting closer to examine each panel, they could see the painted images of hominids. Dr. Fouya gestured, indicating he believed this was a chronological series of events in this culture. Using their respective implants to record all they saw, the team gathered as much information as they could in the time allotted. Dr. Fouya had brought a small chisel which he used to chip some of the walls off to do further analysis.

When the oxygen levels were getting low, Tanita signaled it was time for them to leave, and they retraced their journey out of the hall into the passageway and out of the cave's opening. Getting back in the submersible, Tanita took them back to the surface alongside *Penelope*.

"What do you think, Dr. Fouya?" asked Asher as Tomus and Renard were stripping out of their respective body suits. "Is this legitimate?"

"Too early to tell Mr. Weinbein," was the professor's reply. "I need to do more analysis on the murals and due some radiological dating of this material."

"In my opinion, Asher, I think it is genuine," said Tomus. "On the trip over here, I have been studying the geological records for Kanva, and data show that the ocean has been steadily encroaching on shorelines. It is due to the increasing stellar activity of Trappist-1. At the rate measured, that cave was on land some four million E-yr. ago. The fact it has been submerged for so long makes it almost impossible for that art to be contemporary."

"So you are saying, Tomus," Tanita interjected, "there was a civilization on Kanva that predates Earth's human presence."

"I would feign to make that statement definite, but yes, Grands, it is looking that way."

Chapter 10.

Back at the palace, Renard used the mass spectrometer provided to date the chips he had extracted from the cave and confirmed Tomus' prediction that the material was over four million years old. The ideograms on the walls told a story, according to the professor, of a culture that worshipped the primary star as a deity. How this civilization came to Kanva, what happened to it, and what is its connection to humans on Earth could not be ascertained from the evidence.

Projecting the images from his implants on a screen, Tomus tried to create a narrative.

"Professor, it appears to me that just as on Earth, these people began as hunter-gatherers, as seen by the hunting scenes, then they later evolved into a farming community as pictured in the planting and harvesting images. The question is, why is there no evidence of further development along the lines of Earth's civilization."

"Something catastrophic might have occurred, such as a collision from a meteor or comet, a sudden climate change, or a plague that wiped them all out."

"It would be interesting, Professor, to search other places on the planet to see if there is evidence of other activity."

"Yes, that needs to be done, my boy."

"I know this is not scientific, but the encroachment of the ocean onto the land reminds me of the old Western Bible story of the Flood that the Hebrew God evoked to destroy most living things on Earth. If your recall, Professor, one human family survived with a selection of animals and plants. Perhaps on Kanva, there were no hominid survivors of its flood."

"My dear, Tomus, you are now dealing in myth and fantasy, but I agree that this will be an attractive way to publicize our findings."

"My mother wants us to use discretion in announcing this discovery. I think we need to trust her to find the best way to communicate what has been found."

"This is the achievement of my lifetime. Do you really expect me to sit on this while politicians like your mother, excuse my reference, figure things out?"

"Professor Fouya, you promised confidentiality when you agreed to accompany me to Kanva."

"First, you steal the woman I was interested in sleeping with, and then you order me to hold back on my greatest achievement. My dear boy, I know you like art history and old literature. Here is a reference to chew on. I realize now that by coming with you, I made a Faustian bargain with me being the good doctor and you the Devil."

Chapter 11.

Tomus and Renard presented their preliminary analysis to Alyie and Revita as they met in a conference room at the palace.

"The radiological dating provides strong evidence, and your chronological scenario makes sense," Alyie commented, "but as you recommend, I would prefer to have a complete picture of the history of this civilization before we announce this discovery."

"It seems we need to have an archeological team search the planet for further evidence of this ancient culture," Revita said.

"Yes, I agree," Alyie replied, "but I would prefer it be done quietly. Once we have the full picture of the rise and fall of this ancient society, we can go public. Tomus, can you commission a team to come to Kanva?"

"It will take a little deception, Mother. I would suggest Queen Revita formally petition for a team to come to Kanva and consult on areas of the planet that would serve as future cultural exhibition sites. I will give my approval and forward this request to the Ministry of Arts and Culture. Once the team arrives, we can apprise them of the real purpose of their visit and provide them with the necessary equipment."

"They will need remote sensors such as LIDAR (Light Detection and Ranging) to sweep the entire planet for hidden archeological evidence," said Renard Fouya.

"That can be done," Queen Revita answered. "I will have my engineers begin fabricating the instruments to your specifications, Professor Fouya, immediately."

"Once the team has completed the survey, we will confer and find the most effective means to inform Sidereal Federation Headquarters of what we know," Alyie concluded.

“Excuse my impatience, High Governor and Your Majesty, but I hope there is no delay or suppression of information in communicating this discovery. I am anxious to publish my findings.”

“I understand your eagerness Dr. Fouya,” Alyie responded. “However, a discovery such as this appears to have implications far beyond its scholarly value. You need to leave this to us.”

The team arrived in Kanva about three E-wk. later and were told by Tomus and Renard their real mission. Using airjets equipped with LIDAR sensors, they swept the planet for four E-d. looking for signs of non-natural structures both beneath the land and sea and on both the bright and dark sides of the planet. After completing their survey, the evidence indicated there were five constructed structures that looked either like homes, cemeteries, and religious pyramids. Three of these were under the ocean, while two were on land, buried underneath hills.

Presenting their results at a gathering in the palace, the team concluded that yes, indeed, there was a pre-habitation civilization on Kanva. However, because of a planet-wide flood, the civilization did not survive.

“Any speculation on what caused the flood?” Asher asked.

“My guess,” Jaynis responded, “was that some change in orbital period occurred that precipitated the flooding. Look, the planet has a synchronous four E-d. rotation. Half of it is in perpetual daylight, and the other half is in darkness. This may not have been the case four million E-yr. ago. Perhaps, the planet was forced by instability or some external event to drift closer to Trappist-1. This proximity caused its rotation to lock in synch with its orbital period. This change had to have major environmental effects on its surface, including producing the gigantic Andaman Ocean it now has.”

“The original Kanvaians weren’t able to survive this disruption,” Tanita surmised. “This demonstrates how tenuous life is on a gravitational surface in orbit around another object. We, humans, were lucky to escape a couple of climatic near misses."

"My question, honored Sirs and Madams, is when can I publish my findings?" asked an exasperated Dr. Fouya.

"Dr. Fouya, we appreciate your desire to announce this discovery whose priority actually goes to Commander Tanita Starbright and Mr. Asher Weinbein," replied Alyie. "However, before you submit your research, we would like to alert the High Governor and the Vice High Governor of what has been ascertained by all involved. Then, I imagine there will be a Federation-wide hyperlink news conference of which you will, of course, be invited to participate as an expert consultant."

"And how long will that take to schedule, Madam High Governor?"

"That is out of my hands, Dr. Fouya. That will be up to the present High Governor and his Board to decide."

"This is too much," said a red-faced Dr. Fouya as he excused himself.

Later that day, the 30-odd worlds of the Sidereal Federation were buzzing with news that evidence of an ancient civilization that predated humans on Earth had been discovered on Kanva.

Interlude

In the next three E-m., the Sidereal Federation was thrown into turmoil of to the extent that hadn't been seen since the invasion of the B?alamians some 13 E-yr. ago. All manner of rumors circulated throughout Human Space as to what the discovery of a prehistoric civilization on Kanva meant. Were these ancient hominids precursors to Earth's ancestors? Did they represent a new branch of Homo sapiens? Was there a super race of beings that pollinated planets such as Earth and Kanva with human seed? Or was this some elaborate hoax to create rifts in the Federation?

The clamor for answers grew louder and sectors of the Federation, just as Alyie feared, starting questioned the legitimacy of Earth as the center of Human Space. On Nespar and Ergo there were incipient movements to separate themselves from the Federation and become independent seeing no longer an imperative to pledge allegiance to an Earth-based culture.

The upper hierarchy of the Sidereal Federation took a very defensive view of these events. High Governor Kamesh Thakur assumed these revelations were part of a conspiracy against his rule. A longtime member of the now defunct High Council, he had served for a time as Alyie Starstriker's Deputy High Governor until she reformed the governing structure. Thakur had always harbored resentment and jealously over Alyie's meteoric rise up the Federation ladder and bristled at comparisons between her charismatic leadership and his more procedural methodologies. Objections raised as to his conduct as High Governor festered inside him a hatred that reached a boiling point. Informing Vice High Governor Bristol Menhavi of his unfounded suspicions, Thakur demanded a formal inquiry into the discovery of the artifacts on Kanva, questioning the motives of the House of Jadeite and former High Governor Starstriker. The Vice High Governor, who had served under now General Reg Jones in both his Panthera unit and during the General's term as Vice High Governor, was at first reluctant to take on this assignment, but to mollify the High Governor's increasing paranoia, she assented to proceed with this inquiry.

Not satisfied with Menhavi's lack of enthusiasm with his plan to deflect blame for the crisis on Alyie Starstriker and her associates on Kanva, High Governor Thakur began plotting an alternative means of ending this discord and at the same time finally extracting himself from Alyie Starstriker's umbra.

Chapter 12.

Because of his breach of confidentiality, Renard Fouya had been removed from the investigation on Kanva and had returned to Earth. Tanita and Asher also left Kanva and returned to Cognito. Tomus, with his team of archeologists, had remained, receiving an indefinite extension as they gathered more evidence of the archaic culture. Melissa Turnberry still served as his aide, and the two became quite friendly. A typical day would find Melissa flying Tomus in an airjet over an identified site or traversing on foot a dense forest or hillside looking for remnants of the foregone culture.

Dressed in short khaki pants, a buttoned shirt, floppy hat, and hiking boots, the lanky Tomus seemed ill-fitted at first for fieldwork. However, under Melissa's tutelage, he gained confidence and skill, although Tomus had difficulty keeping up with her demanding pace.

As Melissa was striding up a hill to get a Kanvaian condor's view of the terrain, Tomus, somewhat further down the slope, shouted, "Slow down, Mel. I am not equipped with goat-like genes to bound up mountains like you."

Melissa spun around, flashing her green eyes, smiling broadly with hands on her hips, replying, "Oh, Starbrook, you have your family legacy to live up to. Don't let them down."

Looking at the nubile figure of this young woman standing above him dressed in a tan outfit of a button-down short-sleeved shirt, matching tight breeches, tall brown boots, and a wide-brimmed hat, Tomus couldn't help wish that if she took an interest in his type, he would be glad to reciprocate in turn.

Mutually consenting to rest at the summit of the hill, the two sat down and sipped cool drinks out of the canteens they had brought.

"I didn't think I would say this, but I really like being outside searching firsthand for artifacts. Most of my life, I have spent indoors combing hyperlink sites and digging through dusty archives. It is great to experience discovery firsthand."

"I am going to make you a first-rate field researcher, if nothing else."

"I am glad you took this assignment looking after me. I am sure this was not high on your list of duties to perform in service to the Queen."

"It is not the assignment but the person that makes this enjoyable. Cheers to you, Tomus."

"Cheers to both of us," Tomus replied as they clanked each other's cups.

"On a more somber note, I understand there has been some trouble with our work here and your mother's part in this investigation."

"Oh, it is the usual politics of the Federation. When my mother was High Governor, her leadership was so admired that no one dared oppose her rule. Now that we have a less-than-stellar person in that position, questions are raised about the direction of the Federation. It seems to me that the discovery we have made on Kanva has opened fissures in the unity of the Federation that the current leadership has been unable to close. Their scrutiny of our efforts here and Mother's motives for announcing it are an attempt to scapegoat the problem onto her."

"All I can say to this is that I am sure Queen Revita will stand strongly behind your mother, and you have, for what it's worth my full support."

"Thank you, Mel. That means a lot to me."

Chapter 13.

An E-wk. later, when Jaynis was conferring by video hyperlink with a colleague concerning gravitational quakes, Alyie burst into his study, clearly disturbed.

"It seems I have become the prime laser target for the High Governor. I just received a communique from the Vice Governor's office that I am requested to appear at Federation HQ in Teergarden to answer questions concerning my role in the discovery of the artifacts here. A spacecruiser is coming to Kanva to retrieve me."

"This is outrageous! Here you are trying to help the Federation, and they suspect you of being a conspirator. The Queen needs to know about this."

After messaging Revita and Tomus, Jaynis had Alyie sit beside him while he uncorked a bottle of tawny port wine to settle her nerves. When the Queen was notified of the summons, she rushed to be with the couple.

"How dare they! I have sovereignty over this planet, and as long as you and Jaynis remain here as my guests, you are technically my subjects. You cannot be extradited without formal charges made."

"Actually, the summons is made as a request, not an order," Alyie replied.

"A request that you can respectfully decline," said Revita.

"Oh, Revi. I fear my refusal will only bring stronger measures to bear."

"Then my planet will bear them. If you think I am going to let these dolhogs come here and take my best friend and the savior of the Federation without a fight, you don't know me."

“Let me go. I have the truth on my side. Once I make the case at HQ, it will be clear there was nothing nefarious concerning this discovery. I don’t want any of this to result in open warfare and loss of lives.”

By this time, Tomus, accompanied by Melissa, had joined the threesome. Being apprised of the situation, Tomus volunteered to accompany Alyie on her trip to Teegarden to serve as an expert witness.

“Thank you, Tomus, but the message states I am to come alone, carrying no weapons nor devices that can be used tactically. Besides, being my son, your testimony on my behalf would, I am sure, by those in power, be viewed as biased.”

“Well, the least I can do is send you all the data Melissa, and I have collected in our surveys. That should be more than enough to convince them of the authenticity of this find.”

“I fear,” Jaynis remarked, “that data and careful analysis will not be sufficient to convince these people. I think they have already created a narrative that fits their political agenda.”

“Then,” Tomus said, with his blue eyes flashing, “we’ll need to find another means.”

Chapter 14.

The spacecriuser, *Hesperus*, arrived in Kanva two E-wk. later. After a one-day stopover, Alyie presented herself to the six-person crew to board the return flight to Teegarden. She was dressed in her Admiral's uniform of dress whites: double-breasted jacket, crisp button-down shirt, slacks, gloves, and cap. The white color palate was only broken by her black knotted tie and tall black boots. She did feel somewhat undressed without her military belt, laser pistol, and holster. Jointly saluted by *Hesperus*' Commander, Singha Khan, and Co-Commander Grazia Alvarez, she returned the salutation.

"Greetings, Admiral and High Governor Starstriker," announced Singha Khan. "I know you are coming with us under some duress, but be assured, we have the utmost respect for you. Everyone on board the *Hesperus* has pledged to make this trip as pleasant as possible."

"Thank you, Captain Khan. I look forward to flying with you, especially on this new class of spacecruiser."

At the spaceport to see Alyie off were Jaynis, Revita, Tomus, and Melissa.

"Mom, remember to present the results in the order I sent to you. The evidence speaks for itself, but if you need any additional proof, contact me, and I will send it."

"Thank you, Tomus. Also, thank you, Melissa, for being such a devoted companion to my son. Your presence has changed him in many ways, all for the better."

"It has been a symbiotic companionship, Madam High Governor. He has altered my perspective as well."

Revita came forward to hug Alyie, saying with tears in her eyes, "I know you are only going to save Kanva from a war. I should have mounted a siege on your chateau, preventing you from leaving."

“What kind of friend would that be, stopping me from leaving of my own free will.”

“A friend who loves you more than anyone she has ever known.”

Finally, Jaynis wrapped his arms around Alyie, holding her tight.

“I know you well enough that this is not goodbye. This will not stand. Remember, you are not alone.”

“Be well,” Alyie replied before they kissed long and hard.

Chapter 15.

The trip aboard went smoothly for the first few E-d. The acceleration of the ship in interstellar space gave the sensation of gravity. Alyie was accorded a private cabin on the spacecraft. She took her meals in what served as the mess hall, and although the situation was somewhat awkward, she enjoyed the camaraderie with the crew members, who Alyie noted were particularly young and relatively inexperienced.

Second Lieutenant Emilie Aubere, a petite young woman with curly brown hair and almond-shaped brown eyes, was enthralled as Alyie regaled all sitting with tales of her missions both as a Valkyrie Commander and Fleet Admiral.

"Admiral Starstriker or I mean Madam High Governor," Emilie stammered, confused as to how Alyie should be addressed.

"Let's make it simple, Lieutenant. Since I am a private citizen, just call me Alyie."

"Yes, Ma'am. What I wanted to say is I have met your daughter, Lieutenant Starbrook, on New Cedonia. She is an amazing pilot. We are all in awe at her flying ability."

"Yes, my husband and I are very proud of Tessa, although sometimes she takes chances we do not quite approve of."

"As soon as she commands her own Fighter Squadron, I plan to petition to serve under her."

"That's all nice and good, Lieutenant, but for now, you should learn all you can serving under Captain Khan and Lieutenant Alvarez. They are very capable pilots."

"Just being on the same ship as you is the highlight of my career."

"You are very kind," said Alyie, now feeling like some ancient relic. "But let's both hope there are many other personal highlights in your future."

Chapter 16.

Shaken awake in her cabin, Alyie was slammed hard against the side of her bed by vibrations, breaking her restraints. Floating now due to the apparent loss of simulated gravity, she heard the screeching sound of an alarm indicating an emergency aboard *Hesperus*. Managing despite the loss of gravity, Alyie put on her breeches and pulled up her boots while still clad in her undershirt, all the while feeling a sharp pain in her right side. Sensing a loss of air, she quickly attached the oxygen canister around her neck and its tubes in her nostrils. Opening the door of her compartment, Alyie was engulfed in smoke. Making her way handhold by handhold to the command deck, Alyie could hear cries of anguish ahead. As she reached the entrance to the bridge, she was met by Lieutenant Aubere, who seemed in a state of shock.

"We are under attack! The command deck has been hit badly. They are both in there. It's awful."

"Lieutenant, compose yourself. Remember your training. I see the force shield is not turned on. Do that immediately and give me an assessment of the damage to the rest of the ship."

"Yes, Ma'am."

Alyie then proceeded to climb onto the bridge where through the fire and smoke, she could see the lifeless bodies of Captain Khan and Lieutenant Alvarez splayed in their seats, burnt almost beyond recognition by the laser cannon that bore through the observation deck. Realizing there was nothing more to be done here, Alyie retreated back to the main part of the ship where the four remaining crew members were in a state of near paralysis, awaiting her instructions.

"Seal Command Deck A off from the rest of the ship immediately."

“Yes, Ma’am,” said Ensign Bryan Doherty, a young brown-haired pink-skinned man of sturdy build and brown eyes.

“What is your assessment of the damage, Lieutenant Aubere?”

“The bow has suffered a structural breach. There is also 30% power loss to our propulsion, but life support systems are still operational.”

“Who and where are our attackers?”

“Sensors indicate three fighter class ships are re-grouping some 1000 km. away for a probable second assault. The ship’s log indicates they came within 1 km before opening fire. From readings on weapons systems, the commanders were able to fire a round back before they were overwhelmed. Apparently, this all happened near the hyperknot that would take us out of the Trappist-1 system.”

“Why were these assailants able to approach *Hesperus* so closely without the commanders activating the force shield or confronting them?”

“That is unclear, Ma’am. There was some communication between our command deck and one of the fighters, but due to the damage, I have been unable to decode the logs.”

“We will piece this puzzle out later. More urgent matters concern us. You and I, Lieutenant, are now going to pilot the ship from Command Deck B in the stern. The others, please take your battle stations. We are going to repay these renegades for this treacherous assault.”

When Lieutenant Aubere and Alyie made their way to the stern, the former used her activation code to activate the alternative bridge. Once Command Deck B was operational, Alyie gingerly climbed into the pilot seat with Emilie next to her.

“Madam High Governor, I see you are holding your side. Are you wounded?”

“I banged myself when the attack began. The ship’s autodoc can take care of my injury when this is over.”

Alyie then used the com system to alert Bryan Doherty. “Let me know when the weapon systems are fully charged, Ensign.”

“They are ready to fire, Ma’am.”

Turning the *Hesperus* fully around so that Command Deck B was now facing in the direction of the attackers, Alyie thrust the ship directly towards them, effectively restoring the sense of gravity.

“When I give the command, Emilie, drop the shield and, Bryan, on the count of two, open fire at targets.”

The fighters spread themselves out as the *Hesperus* bore toward them to attack from different directions. But Alyie knew this new class of spacecruiser which she helped design, could lock on targets from its four laser cannons in a full 360° panorama.

The shields held on the *Hesperus* as the fighters renewed their attack. When she deemed they were close enough, Alyie gave the order for Lieutenant Aubere to drop the force shield, and almost immediately after, Ensign Doherty ordered the weapons to fire. Scoring direct hits on two of the fighters and a partial hit on the third, the assault continued until the ships that suffered direct hits were engulfed in flames. Alyie ordered the barrage to cease on the third ship, hoping to recover some prisoners. As she groaned from the increasing pain in her side, Alyie pondered the motive for this unprovoked attack.

(To be continued in Part II.)

The Images on the Wall – Part II.

On a distant day, when his country calls, the king will return. His lost realm will be regained, and his shattered dream of an ideal world will, at last, be realized.

King Arthur's Modern Return, ed. by Debra N. Mancuff

Chapter 1.

Alyie sat in the Commander's chair on Deck B, grimacing with pain after every breath. After receiving notice of the full restoration of oxygen in the ship, she extracted the tubes from her nostrils, allowing herself to breathe more freely. Alyie knew she had broken a rib from the jolt that awakened her when the *Hesperus* had been attacked. She could feel the bone protruding below her right breast, and a small pool of blood was forming where the skin had been punctured. The nanobots that coursed through her blood stream could mend minor internal injuries but not one of this magnitude. The ship's autodoc was equipped with a full surgical unit that could deal with her injury, but it would require Alyie to be under anesthesia which she didn't want to undergo until two objectives were accomplished. The first was to interrogate any survivors that remained on the one starfighter that hadn't been completely destroyed by *Hesperus*' retaliation. From past experience, Alyie knew it is best to get statements from prisoners as soon as possible rather than wait until later when they may have a chance to create a less truthful narrative of their actions. The second reason to remain conscious was to be able to influence the *Hesperus*' subsequent trajectory. Due to her suspicions regarding the motive for the attack on the reason for the attack, Alyie needed to convince the crew to abort the plan to take her to Teegarden and find a less precarious (for her) destination. She realized, of course, once she was unconscious, the crew might receive orders from HQ to complete the mission, but Alyie needed to make her case as strong as possible before she went under.

"Ensign Doherty, are we receiving any communications from the remaining attack ship?"

"Yes, Ma'am. Only one of the two pilots has survived, and he has offered to surrender."

"Then, I suggest, with Lieutenant Aubere's permission, we send a space pod with two crew members fully armed to retrieve this prisoner."

"Permission granted. You have your orders, Ensign."

"Once he comes on board, I suggest we immediately conduct an interrogation, which, if you indulge me, will take place in my cabin."

Seeing the pain Alyie was in, Emilie Aubere agreed, "Of course, Madam High Governor."

"Also, Emilie, can you assist me in getting out of this chair?"

The Lieutenant carefully helped Alyie out of the Commander's chair by supporting her left side. The ship was accelerating sufficiently to provide the sensation of gravity. Still, Alyie had trouble walking due to the increasing pain from the protruding rib. Her boots dragged across the floor as she fought against collapsing, holding on tight to the much shorter Aubere. Another crew member, Ensign Stanis Aboru, was waiting to help guide Alyie into her bed. Refusing to lie down, Alyie asked Aubere to drape her Admiral's jacket over her shoulders so that the prisoner could not see her injury.

Chapter 2.

After what seemed an agonizing wait, Bryan Doherty brought the captive to Alyie's cabin.

"I am Sidereal Fleet Lieutenant Emilie Aubere, Commander of the *Hesperus*. This is Admiral and former High Governor Starstriker. State your name and rank."

"I am Lieutenant Omar Fansi of the Third Fighter Squadron, Sector IV."

"What caused you to attack a Federation spacecruiser?" asked Aubere.

"We received orders from our section leader Rear Admiral Osaka."

"Did you have those orders verified?"

"Affirmative. We used the high-security channel, and it was confirmed."

"Did this order state the reason for the attack?" Alyie inquired.

"All the order said was that the ship contained hostile cargo."

As she mulled what or whom the term "hostile cargo" referred to, Alyie commented, "You did not give any warning or opportunity for the commanders of the *Hesperus* to surrender. This is strictly against Federation protocols."

"Our orders were to identify ourselves as a Federation Squadron and open fire when in range. The mission had to be accomplished at any cost."

"For now, I have no further questions of this prisoner, Lieutenant Aubere."

"Neither have I," added Emilie. "Ensign Doherty, please put the prisoner in a stasis field."

"There is one more thing," Fansi remarked as he clutched inside his right sleeve with his left hand. Drawing a hidden dagger, he broke free from his handlers, lunged forward, striking Alyie in the stomach, who reacted instinctively by planting the heel of her right boot hard against his groin.

Ensign Doherty and Lieutenant Aubere quickly wrestled the attacker to the ground, only to find that Fansi had turned the blade on himself, slashing his carotid artery. Turning to Alyie, they were surprised to see the dagger had not penetrated her skin.

Shaking off the shock of the attack, Alyie assured Emilie and Bryan that she had suffered no injury from the prisoner. What she didn't reveal is that while getting dressed in her cabin just after the *Hesperus* had come under attack, she had activated the protective invisible mesh suit that Reg Jones had given her while she was High Governor.

As Doherty transported the dead assailant away, Emilie turned to Alyie, clearly concerned.

"Excuse me, Madam High Governor, but you need to get on the autodoc pallet now. I see the blood pool from rib injury is now showing through your jacket."

"I will, Emilie. Thank you so much for letting me participate in the interrogation. It is unfortunate that Fansi killed himself before we found out about the motives for this attack. Perhaps we can find more from his ship's logs. But before I get knocked out, I want to know your plans for the *Hesperus*."

"I am not sure what to do, Alyie. I have reported the attack to Fleet HQ on Teegarden. I will send them this information we have on the treachery. It is hard to know who to trust if what this prisoner says is true."

“What is the status of the ship? Is it capable of making the journey through two hyperknots to HQ?”

“That we have not ascertained. Perhaps we should find a nearby destination to conduct repairs.”

“Emile, in this system is the small planet, Hari. It houses an Exobiological Institute that was once attacked by the B?alamians but has since resumed its operations. I am sure it would be capable of repairing the damage.”

“That is a good idea that I’ll suggest when I hear from Fleet Command. But my first priority is getting you to the autodoc.”

Realizing she had done the best she could to influence Emilie, Alyie asked, “Can you give me assistance? I don’t think I am capable of standing upright.”

Lieutenant Aubere directed a floating stretcher to come into Alyie’s cabin accompanied by First Mate Albus Decker. The two of them gently transferred Alyie to the stretcher, with Decker holding her by her boots and Emilie carefully getting under her shoulders. As Alyie floated through the ship, the remaining crew members, Ensigns Doherty, and Aboru, stood at attention as she passed them by.

Once in the space reserved for the autodoc, Alyie was placed on the operating table. She deactivated her mesh suit just as she was about to receive the infusion that would render her unconscious. As the anesthetic was flowing through her veins, she noticed that Emilie was reacting to a message sent to her implant. Alyie could only hope the faith she put in these crew members was not unwarranted, and when she awoke, she would not be confronted by hostile faces.

Chapter 3.

Alyie was coming in and out of consciousness. She sensed vague visions of figures hovering over her, fearing she was in a Teegarden holding facility. Finally, her vision cleared, and looking down on her was the rugged brown face of Reg Jones.

"Hello. Welcome to the real world," he said in his familiar, husky, deep voice

"Well, you are a sight for sleepy eyes," Alyie responded. "Where am I?"

"You are on, Hari. Your advice to Lieutenant Aubere was heeded."

"How long have I been unconscious, Reg?"

"It has been 10 E-d. Apparently, you had an adverse reaction to the medication, and it put you in a deep sleep. All your vital signs were good, so the medical staff was not unduly worried. It just took time for the drug to wear off."

"That explains my presence, but why are you here, General?"

"That takes some explaining. Let's just say your children were very enterprising in alerting me of your predicament."

Now sitting up and realizing that the fracture in her rib section had healed, Alyie asked for assistance from Reg in standing upright and putting on the slippers provided. But first, an autodoc came forward and gently removed all the intravenous capillary tubes that provided nourishment to Alyie while she was hypnagogic. Although even though while she was unconscious, her muscles were continually stimulated, it took Alyie some time to gain back sufficient strength in her legs. She was clothed only in a bare hospital gown.

“Am I here only temporarily?” Alyie asked, concerned that she would still be sent to Teegarden.

“The starbomber, *Olympiad*, which I command with Admiral Osaka, is here on Hari. We have 1000 strong soldiers on board. No one is going to force you to go to Teegarden.”

“Admiral Osaka? Before he killed himself, the prisoner said that he received orders from Osaka to attack the *Hesperus*. Did you know this?”

“There are no records in the Admiral’s logs concerning communication between him and Lieutenant Fansi. This was a blatant deception.”

“Then who ordered the attack?”

“That’s what your son Tomus and I are trying to determine.”

“Tomus? Where is he?”

“Your son is on a starfighter heading here. His ETA is later today.”

Alyie realizing that a lot had happened since she underwent surgery, asked, “What is Federation HQ saying about all this?”

“They have declared a Code IV Emergency. Suspicion has fallen on the alliance formed between Nespar and Ergo. I will send you the High Governor’s statement.

“Waking” her implant which had been in sleep mode, Alyie “read” the following:

“The unscrupulous attack on Federation ship *Hesperus* carrying former High Governor Starstriker is an act of treason that will not go unpunished. We are thankful that the former High Governor did not suffer major injuries. However, the deaths of Commanders Khan and Alvarez are a tragic loss, and their gallantry will be forever remembered. The perpetrators of this heinous act will be found to receive the Federation's justice."

Vice High Governor Bristol Menhavi has charged me with the investigation. It took a little coaxing to convince her, but she understands the issues, despite the pressure she is under."

"What is my status here, Reg?"

"The Vice High Governor and I had agreed that you will be the Senior Consultant in the investigation. With your vast experience and the fact you were the target of the attack, what better credentials could you have?"

"I have a feeling, Reg; I need to thank you once again for protecting me. I am sure the High Governor would prefer me in stasis."

"Now, how would it look for the High Governor to arrest the person who foiled the plot against a Space Fleet ship? You have now gone from scapegoat to hero in the eyes of the Federation."

"Why do I not feel secure about the sudden change in fortune?"

"Trust me; there are many on your side."

Chapter 4.

As Alyie was about to begin getting dressed, she activated the personal communication link on her implant and was inundated with delayed messages from Jaynis, Tanita, Tessa, Tomus, and Revita, all expressing concern about her wellbeing.

Video hyperlinking her husband first, she found him in his office on Kanva.

"Hello, Jaynis. As you see, I am alive and well," Alyie cheerfully greeted him while dropping her hospital gown to the floor and turning completely with the assistance of her autodoc around to provide a full picture of herself. "Although, I find injuries now take longer to heal than they used to. My skin is getting less pliable."

"Always the exhibitionist, but to my eyes, you look as great as ever. How are they treating you on this base?"

"So far, so good. Reg is here with a starbomber. He indicated our children had something to do with his presence."

"I am still not clear on the circumstances, but it appears after you boarded the *Hesperus*, Tomus contacted Tessa telling her to alert Reg Jones about your extradition to Teegarden. He sensed there might be some foul play."

"Well, 'extradition' is too strong a term; I did, after all, consent to go willingly. However, Tomus was correct about foul play. I understand he is coming to Hari, but in a starfighter?"

"Yes, he was pretty agitated over the attack on your ship, as we all were. He says he has new evidence on the civilization that predates the colonization of Kanva he wants to share with you. As for the starfighter ride, the less I say about it, the better."

"Jaynis, I fear this has gone beyond establishing whether there was a hominid culture on Kanva that predates humans on Earth. We need to find out quickly who ordered the attack on the *Hesperus*. If not, there is potential for a Human Space wide war and the sundering of the Federation."

"I am reading dispatches that HQ is making threats to Nespar and Ergo. The Federation may be gearing up for an attack."

"That is horrendous news. We have to find a way to avert open conflict."

"Do what you can, Alyie, but stop putting yourself in the laser sights. You are no longer High Governor."

"Yes, dear, but let me remind you, I was just enjoying life with my husband on Kanva when this discovery broke. I didn't seek to be involved, and I certainly didn't want to be called to Teegarden for an Inquisition. The attack on the *Hesperus* was not my doing nor the attempt on my life."

"I understand. All I am saying is try to stay safe and let others lead."

"Both of these states I have found to be difficult, but I will try. Love you."

Chapter 5.

Alyie was able to contact her mother and Revita, but Tessa's channel was not accepting input. After donning her now-cleaned uniform, with Reg holding her securely, she stepped outside the recovery room to find a party waiting for her in the lobby of the G. Soffen Exobiological Institute. Present were the remaining crew of *Hesperus*: Emile Aubere, Bryan Doherty, Stanis Aboru, and Albus Decker, all smiling broadly to greet her. Hugging each in turn, Alyie thanked them for attending to her injury and not completing the mission to Teegarden.

"No need to thank us, Madam High Governor," Emilie volunteered. "Just before you were to undergo sedation, I received a secure message from General Jones, countermanding the order to continue to Teegarden. I advised him that we would like to land on Hari to seek repairs, and the General agreed."

"Still, I give you, Bryan, Stanis, and Albus, full credit for responding to the attack. Your actions were exemplary and a testimony to your training and dedication to the Federation. Anything in my power to honor you, I will do."

First Mate Decker spoke up, saying, "Speaking for myself, Madam High Governor, serving under a commander like yourself has been the highlight of my Fleet career. This is all the honor I need."

The other three *Hesperus* members seconded that declaration.

"Okay, enough with adulation. Just grant me one request. Call me Alyie."

"Yes, Madam High Governor," they all chimed while laughing.

Then a white-haired distinguished-looking gentleman in a Fleet Admiral suit, Alyie presumed was Yamuchi Osaka, approached.

“My deep regrets, Madam High Governor,” said the Admiral, bowing deep enough so that Alyie could see his thick white eyebrows. “The misuse of my name to order that attack is treason of the highest order. Rest assured. We are doing everything we can to track down the source of this perfidy.”

“I completely understand, and thank you, Admiral, for bringing *Olympiad* to Hari.”

Autobots were stirring in the background. When they separated, it could be seen that a long dining table was set with a sideboard full of food and beverages.

Reg said to Alyie, “The staff at the Institute has prepared this celebration in honor of your recovery. Please, everyone, let’s join in this happy occasion.”

“All I had was a broken rib. Do I deserve this?”

Alyie, there is institutional memory at work here. They remember you rescuing them from the B?alamians. Any opportunity to honor you is taken. Enjoy.”

During the luncheon, Alyie had a chance to catch up with Reg.

“Is it true that you are now a father?”

“Yes, two E-m. ago. Brianna and I had a baby boy we named James Crosby-Jones; fortunately, he looks more like Brianna than me.”

“I am so happy for both of you. What a dad you will be.”

“You mean to say what an old man I will be to my son. Alyie, I am 65 E-yr.”

“It is not your age that matters, but it’s your heart, Reg. Yours is the strongest I have known.”

“Thank you. How are you and Jaynis doing on Kanva?”

"Everything was delightful until this happened. I feel a little out of sync with all this."

"Really? You are giving a good impression of your old self. Perhaps it is just the stark contrast between your competence and the lack thereof of the present holder of the High Governorship."

"I don't want to be the center of an insurgency. My prime interest in all this is to preserve the unity of the Federation. It is important for me and my children's sake."

"Speaking of who, you will get a chance soon to hear firsthand their take on the attempt on your life."

"I thought just Tomus was coming to Hari."

"Oh, I think I disclosed something prematurely. All I can say is he has a very gifted pilot flying that starfighter."

"Tessa! Don't tell me she has gone AWOL."

"Since you don't want to know, I will not say."

"Oh, Reg. This is awful."

Chapter 6.

Later in the day, Alyie's fears were confirmed. The *Falcon* Squadron fighter *Accipiter* made an abrupt landing on Hari's spaceport. Jumping out of the starfighter was Tomus, accompanied by Tessa, dressed in her blue and black Space Fleet uniform. Taking off her flight helmet and displaying her blondish hair, the only physical feature that prevented her from being taken as a younger version of her mother, Tessa flashed a broad smile. Insisting on meeting her children on the enclosed landing port even though the autodoc assigned to her advised against strenuous activities, Alyie stood defiant in front of the spacecraft, her boots firmly planted and her arms tucked around her body to control her fury.

"What are you doing here, Tessa? Do you realize what you have done to your career?"

"Happy to see you are well and back to your old self, Mom. No need to thank me for bringing Tomus along."

Turning her ire to her son, "How could you allow your sister to go AWOL? Don't you know she will be court marshaled for this?"

"Mother, since when have I been able to control Tessa. When she found out about the attack on you, she just flew to Kanva and picked me up."

"I don't know what to say. This is a family disgrace."

"Mother, I have sworn loyalty to the Fleet. But, do you really think I was going to stay with my squadron after I found out that members of the Federation may have tried to kill you? Family always comes first."

Alyie's anger abated as she embraced the twins with tears, "I don't know how we will fix this, but there must be a way."

"Excuse my intrusion on the privacy of your family," Admiral Osaka said after quietly coming up behind Alyie and the twins, "I believe we have found a solution."

"Evidence shows I sent an order to Tessa's commander on New Cedonia requesting her involvement in the investigation of the attack on *Hesperus*. Apparently, Tessa received the communication, but Fleet Command on New Cedonia was unable to find it until just now. It is amazing, even though we live in the 26th century, how technology can fail us."

Alyie smiled through her tears, mouthing a silent thank you and then saying, "It is also amazing how we can tweak technology to our advantage."

"There you go, Mom. See, I am fully authorized to protect you."

With Tessa assisting her mother, who was still slightly weak in the legs from her confinement, the four of them walked back to the Institute proper. Inside, they were met by Reg, who guided them into a conference room.

"Tomus and Tessa, welcome. I am sorry for not allowing you time to decompress from your trip, but there is new evidence that we have just received that I want to share with all of you. Admiral Osaka, please."

"My staff has been able to trace the order given to the commander of that fighter group that attacked the *Hesperus* to the Security and Safety Office on Teegarden. Apparently, someone operating there was able to break into my secure channel and use my authorization to send the attack order."

"That means," Tomus interjected, "the treachery began at HQ."

"Do we know the identity of this traitor?" Tessa inquired.

"Tessa, this is where it gets opaque," Reg replied. "At the time the false order was sent, the duty officer was listed as Second Lieutenant

Camilla Fuselli. We sent a 'Person of Interest' order for her detention, but we have been informed by S&S that she is missing, having not reported to work since the attack order was issued."

"She should be not difficult to locate. All Federation employees can be tracked by their implants," Tomus surmised.

"Lieutenant Fuselli's implant indicates she took a spacecruiser out of Teegarden and traveled to Ergo, her home planet," said Reg.

"Ergo?" Alyie gasped. "Once the High Governor finds out about her destination and affiliation, he will accuse its government of plotting the crime."

"I am sorry, Madam, but he already knows," said Yamuchi Osaka, shaking his head. "Apparently, the Security and Safety Office does not keep internal secrets well. The High Governor just announced an extradition order for Lieutenant Fuselli and is threatening an attack on Ergo if its ruling guild does not turn her over."

"Admiral, how has Ergo responded to this order?" Alyie asked.

"I just read a communique from Ergo stating that no one identified as Lieutenant Camilla Fuselli has traveled to their planet in the last three E-wk. They are claiming that the Federation is using this as a pretense to start a war."

"Then we need to work with all our resources to find a way to avert this catastrophe," Alyie replied. "I don't want to sound over dramatic, but it appears the fate of the Federation rests in our hands."

"Yes, Madam High Governor," replied Admiral Osaka, while Reg and the twins nodded their heads in agreement.

Chapter 7.

After a short break for the twins to settle in on Hari and refresh themselves, Reg Jones called another meeting. Included were Alyie, Admiral Osaka, Tomus, Tessa, and the crew members of *Hesperus*. (Emilie was beside herself to be in the company of Tessa and arranged to sit next to her.) Alyie asked Reg if Jaynis could join the meeting on a secure video hyperlink which he assented to.

Reg began. "I called us together as the working group to solve what now is becoming a Federation-wide crisis. I may be the highest-ranking Federation officer among us, but we all know who our leader is. Alyie, what are your thoughts on how we should proceed?"

"Reg, are you sure about this?"

"It is certainly the way I feel. Is there any disagreement?"

With no one raising an objection, Reg declared, "If you will, Madam High Governor."

"Our first step, as I see it, is to ascertain the whereabouts of Lieutenant Fuselli. Did she ever leave Teegarden, and if not, why is S&S claiming her implant was aboard a ship flying to Ergo?"

"We don't even know that Lieutenant Fuselli was the person sending the attack order," Jaynis commented. "All we can say at the moment is that her name was on the duty officer records when the command was sent. There appear multiple layers to this deception. Reminds me of an ancient Matryoshka doll set, in which each doll has smaller ones nested inside."

"That's why we need a multi-pronged approach," Alyie replied. "One team needs to be sent to Teegarden tracing Camilla's actions up to, through, and after the fateful message was sent. Another team will head to Ergo to locate the implant, which I fear has been removed from Lieutenant Fuselli. Are we in agreement?"

All present nodded yes.

"Lieutenant Aubere, is the *Hesperus* fit to return to Teegarden?"

"Yes, Madam High Governor, all the repairs have been completed. She is fully spaceworthy."

"Then, I request that you and any of your crew members who wish to join go, with the mission, to track the movements of Lieutenant Fuselli on HQ. I am fully aware of what I am asking you to do is well beyond your responsibilities and carries the potential for significant risk, but we absolutely need to know this officer's whereabouts and her current condition."

"If this is any help in making up your mind," Reg interjected, "as head of the investigation of the attack on the *Hesperus*, I give you the authorization to conduct an inquiry into Lieutenant Camilla Fuselli using all resources available at the Security and Safety Office."

"Yes, Sir," responded in synch Lieutenant Aubere and the three other crew members.

"Tomus and Tessa, I want you to go to Ergo. Over the years, I have developed some goodwill with the ruling trade guild. Their current leader, Gianni Vincenti, and I have had a long, mostly friendly history. I would go myself, but I fear my presence would only exacerbate the tensions between Ergo and Federation HQ. I have confidence that Gianni would welcome a visit from my children, especially if they help avoid a dreadful war."

"Great, Mom, I mean Madam High Governor," said Tessa. "As soon as *Accipiter* is checked and refueled, we leave."

"Tomus, I was told by your father you have new evidence about the find on Kanva. Would you like to share it with us?"

"Yes, Mother, I do, but with all that is going on and the sense of urgency we are under, I feel it can wait for a less turbulent time."

OK, I name the team going to Teegarden, *The Seekers*, and those going to Ergo, *The Finders*. We older folks staying on Hari are *The Collectors*. We all have our assignments; let's do them.

As the meeting was breaking up, Emilie could not resist speaking to Tessa.

"I don't know if you remember me, but we met at Fleet Academy."

"Of course I do; you were in the class behind me. Congratulations on your Command of the *Hesperus*."

"As you know, it was under unfortunate circumstances, and I realize it is only temporary."

"Still, you have impressed my mother with your leadership. Your future is bright."

"What I really want is to serve with you when you command a Fighter Squadron."

"Consider it done. I think we will make a great team."

"Thank you so much, Tessa."

Chapter 8.

One E-d. later, both *The Finders* and *The Seekers* were preparing to launch. Alyie stood with the twins in front of *Accipiter* as they were about to enter its cockpit.

"You know your mission, don't you?"

"Yes, Mom," said Tessa. "Meet with Vincenti and find the implant. Simple stuff."

"Keep in contact with me. If I get word of an invasion of Ergo by Space Command, you are to leave immediately. I don't want both of you to be held hostage."

"Yes, Momma. We'll high tail it out of there like a bunch of scared Texas jackrabbits," replied Tomus in an effort to affect an old Earth drawl.

"You have changed, no doubt. I don't recall that sense of humor before, and under different circumstances, I might appreciate it."

"Bye, Mom," both twins sang in unison.

While Alyie was with Tomus and Tessa, Reg was giving his final instructions to the crew of *Hesperus*.

"Lieutenant Aubere, I am sending you the contact information of people I trust at the Security and Safety Office. If you meet any resistance there, present the authorization I gave you to conduct this inquiry. Remember you have the flight records from Lieutenant Fansi's fighter. It can serve as entry into trying to ascertain who really sent the attack order. Godspeed to all of you."

"Yes, Sir," responded Emilie.

After both spacecraft left the hangar and began their separate missions, Reg joined Alyie in a walk outside the Institute. Donning full space suits with oxygen tubes, canisters, and helmets, Reg held

Alyie lightly by the arm as they walked around the periphery of the base.

Confronting the inhumanness of space, Alyie marveled at the disparity between the interior of the Institute with its sleek design, soft lighting, oxygenated air, comfortable temperature, and a profusion of blooming flowers and plants with the utter emptiness all around her. How humans have learned to survive in the Universe has been a testimony to their ingenuity and determination, but the comfort of a place like the Institute ignores the few cm. of protection where lies the thin veil that separates humankind from oblivion.

Signaling to Reg she was ready to return back inside. They proceeded to first go into the decompression anteroom, where they stripped out of their space clothes. Reg put on a simple green tee shirt and matching trousers while Alyie stayed in her white undershirt, but she put back on her white breeches and black boots. She felt a chill, perhaps a reminder of the Absolute Zero temperature outside, so she draped her admiral's jacket over her shoulders.

Reg ushered her into a lounge area where they could sit. Offering a cup of hot tea, he could tell Alyie wanted to talk.

"So now we wait," Alyie said with a wane smile, the lines around her green eyes deepening.

"Do you miss being in the center of the action?" Reg asked.

"It seems a few E-wk. ago I was as much in the center as I would ever want to be. By the way, do you remember how resistant I was to wearing that mesh you had designed to protect me? Thank the stars; you convinced me otherwise. Imagine the regrets I would have if I remained vain and obstinate. But I am being silly because there would be no me to have regrets."

"Is there anything you regret?" Reg asked.

Alyie looked into his soft brown eyes while holding his hand. "I don't regret my time with you at all. I do regret being unfaithful to Jaynis. Also, any pain I caused you when we stopped.

"I always knew we were transitory. I am happy everything has worked out with you, Jaynis, and your family. As for me, life is good. Brianna and I are deliriously happy, and now with James, we can't wait to see what will happen next."

"That is great to hear. I am so happy for you, Reg. As far as the larger picture, I fear what will happen next to the Federation. It has lasted almost three E-c., but in Earth's history, even longer-lasting civilizations have fallen. Everything is transitory. Nothing is guaranteed to last. Look at the civilization on Kanva. I am sure they had no expectation of complete destruction."

"None of us do, Alyie. We always live on hope. It's what keeps us going."

"The more I think about those murals on the cave walls, the more I believe they are trying to tell us something. Tomus said he found new evidence. I am really curious about what that is. Maybe it will teach us how to survive this crisis."

"We will survive as we always have with strong leadership bringing humankind together for the common good. That has worked in the past, and with you back at the helm, that will enable us to succeed again."

"Reg, I fear despite our efforts, it will result in failure, but in the back of my mind, I fear even more that we will meet with success. I really don't want an Act II of the Starstriker-Jones High Governorship."

"You may not wish it, but the way I see it, we are the only ones standing between a corrupt Headquarters and a full-scale rebellion."

"There must be others who can step up."

"Perhaps, but Alyie, no one in the entire Sidereal Federation is more respected than you. If you take back the High Governorship, we can avert a civil war."

"I don't want a coup; there must be another way. We must not subvert the rule of law. We need to avoid a conflict but do so within the parameters of the Sidereal Federation Constitution."

"As always, I am here to serve you, my Madam High Governor."

Chapter 9.

Of the two missions, *The Finders* were the first to reach their destination. Gianni Vincenti, the Trade Guild Director, was happy to welcome Tomus and Tessa on Ergo. He bestowed on them all the hospitality of the Vincenti family, insisting they be his guests at the family's spacious compound where also resided his wife Rosetta and their five children.

Gianni was a way silver haired dapper man of trim, medium built, and dark brown eyes. During the welcome dinner in honor of the twin's arrival, Gianni told them stories of their mother's first visit to Ergo over 25 E-yr. ago. He seemed to hint that there was something going on between him and Alyie, but politics got in the way.

"At first, I didn't think I could forgive her for what she did, but in love and war, all is pardoned."

"What did she do, Mr. Vincenti?" asked a curious Tessa.

"She ordered the destruction of our prime research center."

"Oh, that," said Tomus, recalling the secretive use of tachyons to predict in advance economic news.

"Your mother is a difficult woman to forget, but I see she has two beautiful children, and I have my own wife and family. So everything has worked out for the best."

Gianni and Rosetta had three sons and two daughters. From the way Gianna pushed his oldest son, Diego, toward Tessa and his oldest daughter, Concetta, on Tomus, it was clear he thought that if a union between him and their mother didn't work out, perhaps their respective children would find love together. Each of the twins maintained their composure, but the heavy-handedness of Gianni embarrassed his children.

The next E-d. Tomus and Tessa began their mission to locate Lieutenant Fuselli's implant. Gianni sent them to Ergo's Information Center, where they were shepherded by its Director, Mr. Stephen Marco. Camilla Fuselli was indeed from Ergo, and the data showed her mother and father lived in a home on the outskirts of the capital. From recent interviews, neither parent had reported seeing or speaking to her in over an E-m. From the videos, it was clear they were deeply worried about their daughter's disappearance.

"Do you have any tracking of Ms. Fuselli's implant, Mr. Marco?" asked Tomus.

The Director, a pleasant older gentleman, responded by saying, "This is what has confused us. There is no person listed as a passenger on any recent spacecraft coming from Teegarden to Ergo using the name Camilla Fuselli, but we do have a hit on an implant registered in her name on a flight that landed two E-w. ago."

"Do you think she flew under a false name, Mr. Marco?"

"When we tried to identify who was carrying the implant in their head, the device was deactivated."

"Then we need to see a list of the passengers on that flight to find who shut off their implant," Tessa declared. "That must be Camilla."

"That is exactly what we thought to do, Lieutenant Starbrook. However, when we conducted a subsequent search of all passengers on the flight, their implants were fully activated, and none matched the signature of Fuselli's."

After a momentary pause, Tomus spoke, saying, "That, to my mind, means that the implant never left the spaceship that took it to Ergo. Is the ship still in the spaceport, Mr. Marco?"

"Let me check. You may be on to something, Dr. Starbrook."

After accessing the flight arrival and departure schedule on his implant, Mr. Marco, with some glee, expressed, "We're in luck; the spacecruiser Medusa is not scheduled to depart till tomorrow. I will send a nanobot team to search for the implant."

"If you don't mind, we would like to join them," Tomus added. "If we find it, I want to send this news immediately to General Jones and my mother."

Tessa and Tomus boarded an airjet that whisked them, following the security team's airbus, to the spaceport. Medusa was in the large hangar, being checked and fueled for its next trip. The head of the security team presented the search authorization to the Flight Manager and promptly boarded the spacecraft with his fellow members. Dispensing the bug-like nanobots throughout the various sections of the ship, the twins waited outside to get the results of the search.

"I thought Concetta was kind of cute. Did you, Tomus?"

"Tessa, we are not here for that kind of thing."

"As for me, I think Diego is a doll; I wouldn't mind spending more time here getting to know him better."

"Tessa, what are you saying? What about Yong Gwan?"

"Oh, yes. I forgot. It's just we don't get together so often. But your right, we are a couple."

Any further conversation was interrupted when the head of the security team came out of the spacecraft dangling an implant from a pair of tweezers.

Chapter 10.

The Seekers had a more difficult time extracting information upon arrival at Teegarden. At first, the guards at the Security and Safety Office refused them admittance. But once they produced General Reg Jones' authorization, they were granted access to the Departmental Record Office. The Record Officer, First Lieutenant Finneas O'Rourke, who Reg had told them to meet, was at first reluctant to assist them.

"My Department reports directly to the Vice High Governor, and General Jones is not in that position anymore."

"But he has been given authority by Vice High Governor Menhavi to conduct this investigation," Emilie Aubere declared. "We are part of a larger team looking into the attack on the *Hesperus*. Other members include Rear Admiral Osaka, Former High Governor Starstriker, Dr. Jaynis Bottombrook, Dr. Tomus Starbrook, and First Lieutenant Tessa Starbrook."

"Tessa!" exclaimed O'Rourke. "Do you know I used to work with Tessa when she interned for General Jones? She and I were pretty tight," Finneas confided, widely exaggerating his relationship.

Emilie realizing this was an opening, said, "I was just with her on Hari. She told me to give you a big hello. She actually told me to do much more, but we are in a professional setting."

"Tessa did? Wow!"

"Yes, she hopes to see you soon as she is scheduling a trip to HQ."

"Great! I love to show her around. You see, I am now a First Lieutenant, the same rank as Tessa."

"I can see you both have a lot in common and a lot of catching up to do. I'll be sure to tell Tessa how much you have helped us in this mission."

“Okay. Follow me to the Class 10 room.”

Once in the high-security enclosure, O’Rourke using his top-priority clearance to do a search for the movements of Camilla Fuselli around the time of the attack. What he found is that two E-hr. before she supposedly broke into Admiral Osaka’s security network, she had stepped out of the S&S building, which she logged as a scheduled medical appointment. When she returned, video footage showed her dressed in a full-length trench coat with the collars up, hiding her face from the cameras with her implant still active.

“Lieutenant O’Rourke, is there a way to make a body identification of that individual?” asked Ensign Bryan Doherty.

“We do have such technology, but it is only to be used in the most extreme circumstances,” cautioned Finneas. “Privacy laws prevent scanning the public without the most urgent Federation needs.”

“Don’t you think finding the identity of the individual who ordered an attack on a Federation ship, killing its commanders, and trying to assassinate the former High Governor is a top Federation urgency. I am sure General Jones and Tessa would concur,” declared Emilie.

“Yes, of course,” said O’Rourke using the access code to run the identification program on the video image.

After a few E-min., the scan was completed and identified the person as Anya Geranova, Officer, Second Rank at the Security and Safety Office.

“I know her a little,” said Finneas. “She started working here about a year after the new High Governor was elected. She mostly keeps to herself.”

“As I am accessing her background,” said First Mate Albus Decker, “I see she is the daughter of Frida Geranova, a former High

Council member who resigned soon after Alyie Starstriker became High Governor."

"A search of Frida shows she died a few years after her removal, which apparently was under negative circumstances," said Ensign Stanis Aboru.

"I think we may have found a motive for the attack – revenge on Alyie Starstriker," Emilie declared. "I need to get this information to *The Collectors*."

Chapter 11.

"Frida Geranova's daughter!" Alyie cried. "Is this a second-generation payback for my abrupt dismissal of her mother? At the time, Anya must have been no more than 10 or 11 E-yr. old."

"She actually was 11," Jaynis answered. He had recently arrived on Hari, taking the relatively short excursion from Kanva. Huddled together with Alyie, Reg, and Yamuchi, *The Collectors* tried to fit this new piece into the elaborate puzzle of this ever-widening conspiracy.

"Now Anya is 28," Jaynis continued. "Her history shows she received a special appointment to the Security Office from the High Governor himself after working in a Regional Office on Planet Fallow in the Ross System for four years after graduation from technical school. Her elevation to Teegarden was, no doubt, to honor the memory of her mother, who was a close colleague of Kamesh Thakur."

"We need to apprehend her as soon as possible. She is the key to the attack on the *Hesperus* and the attempt on your life, Madam High Governor," said Admiral Osaka.

"But to carry out such an elaborate act of revenge, she must have had assistance," Alyie surmised. "What happened to Camilla Fuselli? Where is she, and how was her implant removed?"

"Also, how was Anya Geranova able to override Admiral Osaka's security code?" wondered Jaynis. "It is my understanding that can only be done at the highest levels of the Federation. Am I correct, Reg?"

"Yes, only the High Governor and Vice High Governor would be able to break into this. This is what I feared all along. That this treachery reaches to the top."

"All we have just said is conjecture," Alyie replied. "We need to interrogate this young woman to find out who aided her."

"The problem is that we are here on Hari, and she is on Teagarden," said Jaynis, stating the obvious. "I don't think, General and Admiral, that sending the starbomber *Olympiad* to the military headquarters of the Federation and telling them to turn over Anya Geranova, who may be acting as an agent for one or both of the heads of the government, would be received kindly."

Reg, who seemed to be wrestling internally with a conflict, resolved to say, "The only person who can legitimately order the arrest of Anya Geranova is Vice High Governor Menhavi. I know this is taking a great risk, but I propose we trust her by sharing the evidence we have collected from Tessa and Tomus' trip to Ergo and what the other team has discovered on Teegarden. I had mentored Bristol Menhavi when she was a Panthera. I have always found her to be upright and loyal to her oath to serve the Federation. My sense is that she is uncomfortable with the High Governor's recent behavior and his hostility towards Nespar and Ergo."

"You realize if she is carrying out Thakur's orders, then telling her what we have discovered exposes our position and places *the Seekers* in great jeopardy on Teegarden," Jaynis replied. "We all will be labeled as traitors."

Take the risk, Reg," Alyie replied. "I trust your judgment of Bristol."

When Reg contacted the Vice High Governor by video hyperlink, he found her in the Command Center on Teegarden HQ. She said she couldn't speak at the moment, but she would contact him on a secure channel when she returned to her office.

A few E-hr. later, Reg received a return call from Bristol Menhavi. Looking at the worried face of this fortyish-year-old woman, Reg could see she was under considerable stress. There were furrows in her brow, her brown eyes were bloodshot, and her normal dark-skinned complexion showed spots of yellow pallor. She was in her full dark green Colonel uniform, but she had unbuttoned the left side of her double-breasted jacket. She sat crouched in a chair with her hands pressing on either side of her face and the heels of her black boots propped up against a small stool.

"Hello, Reg, I am sorry we couldn't talk before, but I was receiving a disturbing call from the High Governor."

"Hello Bristol, are you alright? I hate to say this, but you don't look your best. Have you not been sleeping?"

"Oh, Reg, who should know this better than you; it is the cost of the job."

"Well, I have more problems to dump on you. My investigation of the attack on the *Hesperus* has led back to your neighborhood. We have strong evidence there is a traitor working in the Security and Safety Office on Teegarden.

"I am sending you now all we have acquired. You will see that this person assumed the identity of another security officer, was able to break into Rear Admiral Osaka's top security account, and, in the Admiral's name, ordered the attack and then subsequently attempted assassination of the former High Governor."

"This is incredible; what could be the motive for this?"

"On the surface, it appears to be driven by a long-simmering quest for revenge on the former High Governor for her demotion of the assumed perpetrator's mother some 17 E-yr ago, but I am afraid to say the conspiracy goes much further."

"How so, Reg?"

"Before replacing the duty officer, Lieutenant Camilla Fuselli, at the S&S, the suspect, Anya Geranova, with outside help, we think, captured Fuselli and extracted her implant, which was put on a spacecraft headed for Ergo to look like Fuselli was an agent for that planet. Part of my team has traveled to Ergo and is now in possession of the implant."

"Where are Lieutenant Fuselli and your suspect?"

"We need your help to find them both. I have a team on Teegarden working with your S&S Office. We have located Geranova, but we need you to arrest her. Our hope is through interrogation; she will tell us the whereabouts of Fuselli and the identities of all who conspired with her."

"Yes, I will look at what you have sent and take appropriate action, General."

"I feel I have to tell you, Bristol, this may go very deep. I am putting my trust in you, but aiding in this investigation is not without great risks for all involved."

"Understood, and in the spirit of mutual trust, I am breaking security when I tell you that call from the High Governor was an order to send a starbomber to Ergo. We live in perilous times, old friend."

"God be with you, Bristol."

Chapter 12.

When Reg shared thc news of the starbomber *Dreadnought* heading for Ergo, Alyie, and Jaynis reacted immediately. Contacting the twins, the parents urged them to leave before the starbomber arrived.

"Mom and Dad, Tomus and I have discussed this, and we have decided to stay and help defend Ergo."

"Listen, Tessa, I am your mother, and I am ordering you to return," Alyie demanded.

"Yes, you are my mother, but you are not my commander. I swore an oath to defend the people of the Sidereal Federation when I was commissioned in the Fleet. The people here on Ergo are still members of the Federation, so it is my duty to protect them. Besides, I like them a lot."

"All your mother is saying is that as *The Finders*, you have completed the mission in finding the implant; it is more important to come to Hari with the evidence than to stay on Ergo."

"Nice try, Dad, but let me ask Mother if she was in our place would she desert these people?" Tomus questioned.

"That is irrelevant because this situation is different. What could you both do to stop an attack from a starbomber?"

"No, it is not irrelevant," shot back Tessa. "First of all, you have set the professional standard I live by, and second, with me piloting an ultra-fast ship like *Accipiter*, it can wreak havoc on a slow starbomber such as the *Dreadnought*. Just you watch."

Tomus felt the need to diffuse the situation. "Mom and Dad, there may be a way to defer any hostilities. What I am working on is a strategy to delay any combat until your investigation is complete. Have faith in us; we will do what is best."

“Okay, but don’t take any wild risks,” Jaynis replied as Alyie was too tied up in emotions to speak.

Shortly after, Reg called a quick meeting of *The Collectors*.

“Yamuchi and I have conferred, and after doing the orbital mechanical calculation, we have determined that if we leave right now, the *Olympiad* can reach Ergo space approximately when the *Dreadnought* is nearing the Tau Ceti system. Thus, we will be able to provide a deterrent on any attempt to attack Ergo.”

“Good,” Alyie replied. “I can be ready in 30 E-min.”

“No, Alyie, you’re not going. You and Jaynis are staying here to move forward with the investigation.”

“Perhaps you don’t understand, Reg, because James is so young, but my children are on Ergo. I am not going to sit here like some doddering old woman waiting to hear if they survive or not.”

“Alyie, the image of you as a ‘doddering old woman’ is so incongruous to your nature that if things were not so perilous, I would take time to laugh out loud. No, you are staying here. I hate to remind you, but you have no rank or position in the Federation.”

“Reg, why are you doing this to me?” said Alyie, nearly in tears now.

“I am doing this because it is the right thing to do. I cherish Tessa and Tomus, and you must know that I will do everything in my power to keep them out of harm’s way. But the more important reason for you and Jaynis to stay in Hari is to interrogate Anya Geranova once she is apprehended. If we are to avoid a calamity on Ergo, we need a confession from her. Who better to find out the truth than you and Jaynis?”

“Reg is right, Alyie,” Jaynis concurred. “I trust the kids to take care of themselves and Reg and Yamuchi to provide a buffer to prevent a war. We need to let go.”

“Okay, but this is so hard for me.”

“I know Alyie,” consoled Reg. “I am leaving my Adjunct, Major Joseph Kreiger, behind to assist you. His presence will provide the official imprimatur to your interrogation of Geranova.”

“Reg and Admiral Osaka, please be safe.”

“We will try, Madam High Governor,” replied both Reg and Yamuchi.”

Chapter 13.

It would take about eight E-d. for *Olympiad* to reach the Tau Ceti system by the optimal hyperknot route calculated by Admiral Osaka. In that time, Alyie and Jaynis waited for news of the apprehension of Anya Geranova. It came after two days when Major Kreiger announced Vice High Governor Menhavi informed him of her capture.

"It did not come without incident, Madam High Governor and Dr. Bottombrook. She was about to be abducted in her flat by two assailants when the Safety and Security Task Force came to make the arrest. After a standoff, the assailants surrendered and turned her over to us."

"Who are these abductors?" Jaynis inquired.

"We have ID them, and they are freelance contractors who have a history of malevolence. They were interned in our stasis facility on Teegarden but had their 20 E-yr. sentence recently commuted by High Governor Thakur."

"Major Kreiger, you don't find that strange?"

The Major, who was a studious officer still in his thirties with dark brown hair, brown eyes, and white complexion, held back for a moment before saying, "the more I am involved in this case, the stranger it becomes, Dr. Bottombrook."

"Does the arrest of these individuals help us understand the fate of Lieutenant Fuselli?" Alyie asked.

"I am afraid this news is bad. The two perpetrators, after intensive interrogation, have admitted to abducting Lieutenant Fuselli. They claim that in removing her implant, she suffered a cerebral hemorrhage and died in the autodoc clinic."

"Is there any evidence of this?" Jaynis asked.

"They disposed of her body, but a search of the clinic found blood stains that are clear matches for Camilla Fuselli's DNA."

"Who put them up to this scheme?"

"That's less clear. They claim they were contracted by a source from Earth who we have found using a false identity. However, the S&S team has traced the messages sent to these criminals to Federation HQ in Colorado Springs."

"When can we interview Officer Geranova?" Alyie inquired.

"I have been informed she is being prepared as we speak. You should get the opportunity Madam, soon."

After a 2 E-hr. wait, a sullen, somewhat disheveled young woman with light brown hair and pale blue eyes appeared on the video hyperlink screen.

"Officer Geranova, do you know who I am?" Alyie asked.

"You are the person who destroyed my mother's career and led to her death."

"Yes, I am Alyie Starstriker, and I take full responsibility for dismissing your mother from service on the High Council. We had reached an impasse in which I felt I couldn't govern with her around. I do not take any responsibility for her death, but I am sorry for her loss and the grief it has caused you."

"No, you're not! You were High Governor for 15 years while my mother lived only a few years after in disgrace."

"I can understand your hatred for me and the need to seek retaliation, but do you know the lives you have cost or put in peril in your drive for vengeance?"

"Yes, I am sorry for the deaths of those commanders on the spaceship, but they were in the Fleet and have been trained to accept death."

"But what about Camilla?" asked Alyie. "She was not in the military. Did she deserve to die as well?"

"Camilla dead?" Anya gasped. "I was told no harm would come to her."

"You are either naïve or not telling the truth," Jaynis remarked. "When you were given her implant, what did you think happened to her?"

"I was told she was going to an autodoc clinic for routine examination in which she would undergo sedation. Once she was unconscious, the extraction would take place. After I fulfilled my role, I would hand back the implant to my contacts, who would insert it back into her before she regained consciousness."

"Well, that is not what happened, Anya. The extraction was performed poorly, and Camilla died."

"How do I know you are making this up to trap me?"

"Here is the DNA report, confirming Camilla's presence in the medical clinic, and here is your co-conspirator's confession on what happened to her," Jaynis replied

"Oh, I never meant any harm to her."

"There is more, Anya," Alyie added. "The implant was put aboard a spacecruiser heading to Ergo to make it look like Camilla Fuselli was an agent for that planet's government. The High Governor has sent a starbomber to Ergo, accusing them of engineering the attack on the *Hesperus*. Your desire to kill me has been used as a pawn in a wider scheme that now threatens millions of lives."

"I never knew any of this, I swear," said Anya, now visibly upset.

"What we need to know, Anya is who put you up to this and gave you the ability to break into Admiral Osaka's private network?" asked Jaynis.

"It was High Governor Thakur. He sent me the access code as soon as I replaced Camilla on duty. He said he owed it to me for my mother's support of him over the years. He promised that Camilla would be absolved of any culpability."

At the name of the High Governor, Major Kreiger, Jaynis, and Alyie all sat back in their respective chairs, sensing this inevitability but nonetheless collectively upset over the consequences of its revelation.

Chapter 14.

All the information collected in the entire investigation, including Anya Geranova's confession, was sent to Vice High Governor Menhavi. Bristol thanked them for their service but was circumspect in terms of what would happen next.

Somewhat taken aback by Menhavi's hesitancy to bring charges on the High Governor, Alyie pressed her, "This is a capital crime. It needs to be brought to High Justice Victoria Alana. Once she reviews the evidence, formal charges can be made. If so, ultimately, it will be the High Governor's Board who decides whether the High Governor can continue to serve."

"I know the protocols, Madam, but I do not want to proceed until I am absolutely convinced of his complicity in these crimes. You have to understand the sensitivity of my position."

"Okay, Bristol, but the least you can do is call off the *Dreadnought's* mission to Ergo. Surely, you realize now that the government of Ergo played no part in the attack on the *Hesperus*."

"That would impel me to countermand a direct order from the High Governor. As long as Kamesh Thakur sits in the High Governor's chair, I have no power to intervene."

"Bristol, listen to me. It is not a question of power, but it is the need to do what is right. How do you want to be remembered? As someone who obeys orders causing death and destruction or as a leader who stood up to tyranny."

"Once my office completely reviews the evidence and if the High Justice issues a warrant, I will act, I assure you, Madam."

I fear we can't wait that long," was Alyie's reply. "We have reports that in addition to Ergo, Nespar, and Cognito are mobilizing their defense forces."

Alyie updated Reg by video hyperlink on all that had transpired, including her frustration with Bristol Menhavi's reluctance to act expeditiously on bringing forth the case against Thakur or rescinding the mission to Ergo.

Reg, clearly disappointed, said, "I have always regarded Bristol as a first-class officer, but the only fault I saw was her strict adherence to rules of order. In this case, this rigidity has potentially devastating consequences. Let me try to reason with her."

"My feeling Reg is she doesn't want to be seen as the one to topple Thakur. She'd accept others doing the job, but she does not want to be known as the person, as in the old expression, 'wielding the knife'. I can sympathize with her a little. I, too do not want to be accused of using this crisis as an excuse to return to power. That would only plunge the Federation into a Human Space wide civil war. Maybe there is a way to keep us both out of the laser crosshairs."

"We can hope, Alyie."

"Good luck, Reg."

Chapter 15.

When the *Olympiad* came out of the hyperknot that put it in the Tau Ceti system, Reg tried reaching the Vice High Governor, but he found she was not accepting input from him. Clearly disturbed, he conferred with Admiral Osaka, who resolved to continue to Ergo in the hopes of intercepting the *Dreadnought*.

After two days, the starbomber was sighted on a direct course to Ergo. Osaka increased the thrust on *Olympiad* to place it directly in *Dreadnought*'s path. Once in communication range, Osaka hailed the Commander of the *Dreadnought*, Captain Taneous Taub.

"Captain Taub, this is Yamuchi Osaka aboard *Olympiad.* Please cease your mission. We have proof that Ergo was not responsible for the attack on a Federation ship."

"I cannot do that, Admiral; only a direct order from the High or Vice High Governor can abort this mission."

"Captain Taub, this is General Jones; we have proof the High Governor ordered the attack on the *Hesperus*. He is no longer fit to serve. Please stand down."

"Can't do that, General. And I must inform you, if you do not stand down yourself, you will be considered a belligerent that will force us to take aggressive action."

"Admiral Osaka and I cannot let you fire on innocent civilians. Please don't force us to confront you. Just stand down."

"Commanders of *Olympiad*, your presence is a direct act of aggression against an authorized Federation starship. If you do not stand down in the next 30 E-min., prepare to be attacked."

Admiral Osaka gave the order to activate all force shields, and General Jones rang out the cry, "Battle stations, everyone! Charge all laser cannons and get ready to fire on command."

Just as the waiting period was ending and *Dreadnought* was counting down to launch its missiles, a starfighter was seen by the crew of both spacecraft to fly between the two giant starbombers. Reg quickly identified to be *Accipiter*.

"Oh, Tessa," he said to himself. "You are going to get yourself killed in the barrage. How will I ever be able to face your mother?"

Soon after, a message was sent to both ships from the *Accipiter.*

"Hello. I am Tomus Starbrook flying with my sister, Tessa. Before you begin your onslaught, which will bring shame and disgrace to all involved, please allow me to speak."

When neither of the bristling warships stirred, Tomus continued. "This whole crisis began with the discovery of murals painted on underwater cave walls on Kanva. I have been a co-leader in investigating the civilization that produced them. Once I realized the harm the discovery of them brought to the stability of the Sidereal Federation, I began to doubt whether I should have made their existence known. However, from further investigation, I realized that this civilization has much to teach about our own sustainably. The research team and I initially assumed the annihilation of this pre-historic culture of Kanva was caused by climate change that flooded much of the terrain. However, my co-worker Melissa Turnberry and I have found recent artworks from above-ground sites that indicate the culture was not destroyed by changing environmental effects but by devastating intracultural wars. Please allow me to show this in images."

Beaming photos of the art to the video screens of both starbombers, all aboard saw scenes of carnage and death as those who had secured the high ground defended it against attackers. The images vividly depicted the horror and terror in the faces of the hominoids who fought to the last survivors.

"This was a war of attrition, causing destruction not only to the combatants but to the land itself. In the aftermath, those who remained had not enough resources to survive. The civilization was destroyed by mutual annihilation," pronounced Tomus. "My reason for showing this is to avoid taking the first step that may lead to human destruction."

"Thank you, Dr. Starbrook, for showing us these images," said Captain Taub. "Under different circumstances, they may have been illuminating, but we have our orders."

"Can't you stand down for at least 24 E-hr.?" Tessa exclaimed, breaking into the communication link. "Take the time to assess what you have both seen and decide what best course to take. You have orders sent from far away and some time ago, but the fate of the Federation lies in your decision right here and now."

"Agreed, we will stand down for one Earth rotation period," said Commander Taub.

"We also will cease from any action for at least that time interval," added Admiral Osaka.

Chapter 16.

In the tense period during the truce, Tessa kept *Accipiter* in a direct line between the two starbombers daring them to fire at the small starfighter. In the 22nd hour, suddenly, everyone on the three spacecrafts received a video hyperlink message coming from Vice High Governor Menhavi's Federation-wide broadcast band. However, instead of seeing the image of the Vice High Governor, Alyie Starstriker appeared on the video screen. She stood tall, all dressed in black – turtle neck pullover under a sport coat. Her tight black trousers were tucked inside her tall black boots. Only her flaming red hair upset the black silhouette. Looking grave but commanding, she began,

"Citizens throughout the Sidereal Federation, I am Alyie Starstriker, former High Governor. I am speaking to all of you through the auspices of Vice High Governor Menhavi, who has given me access to her broadcast channel. I bring you news of terrible treachery at the highest level of the Federation in the hope once the citizens of the Federation see the evidence collected, they will demand action and justice. I am a private citizen like most of you, but even outside of the government, we citizens can insist that no one, even the highest, is above the law. Let me now show you what an investigative team has discovered."

Alyie then stepped through all the events starting with the attack on the *Hesperus*. She did so by first projecting images of the dead commanders, Khan and Alvarez then showed the video of the attempt on her life by Lieutenant Fansi. This was followed by a photo of Lieutenant Camilla Fuselli and the arrest shots of her abductors. An image of Tomus was projected, showing him holding Camilla's implant with the Visconti villa on Ergo in the background. Finally, Anya Geranova's confession was screened, where she identified High Governor Thakur as the person who gave her the access code to order the attack.

Allowing time for the montage to sink into the minds of everyone receiving the broadcast, Alyie waited until resuming her presentation.

"What we are asking is that all preparations for hostilities cease until justice is served. Let the system work in the name of peace and unity. To those aboard the starbombers *Dreadnought* and *Olympiad*, you heard what my son, Tomus, said about the civilization on Kanva. Don't let us fall into the dreadful trap that culture descended into in choosing hasty violence to settle disputes when reason, although sometimes agonizingly slow, can prevail. I thank the Vice High Governor again for giving me this opportunity and my fellow citizens for listening to my plea."

Epilogue

Alyie had very mixed emotions. She did not anticipate returning to Earth so soon after her term as High Governor had ended. In the past two and half E-yr., Federation Headquarters in Colorado Springs had undergone some changes. The environs seemed less accommodating as there were sentinel monitors everywhere. Yet the trees she ordered planted down the canyon leading the complex of buildings had grown to nearly full height, and there were graceful statutes she had commissioned of human heroes in art and science that surrounded the courtyard where the ceremo ceremony was to take place. Alyie had arrived with a delegation that consisted of her family: Jaynis, Tomus, Tessa, and Tanita. Accompanying them were Melissa Turnberry, Yong Gwan, and Asher Weinbein. Alison Bernewithy, the system's Regional Governor, also was in attendance.

Presiding over what would soon take place on the dais in the center was Vice High Governor Bristol Menhavi. Standing beside her was General Reg Jones, Admiral Yamuchi Osaka, Board member Rufus Seul and High Justice Victoria Alana. In the aftermath of Alyie's broadcast, an informal Federation-wide truce was reached. When High Justice Victoria Alana brought charges against High Governor Thakur, he resisted at first, claiming Executive Privilege. When public sentiment reached a near-breaking point, even Thakur's High Governor's Board turned against him. He attempted to resist removal and subsequent arrest by ordering his Elite Guards to escort him to safety, but most had served under Reg Jones and stayed loyal to the Vice High Governor Menhavi. Turned over to the High Justice, Thakur was now in a Federation stasis facility on Teegarden awaiting trial on charges of treason and conspiracy to commit murder.

When everyone was settled with the ceremony broadcast over the entire Federation, Vice High Governor Menhavi began,

“Honored guests and citizens of the Sidereal Federation, I speak to you today to bring some closure to the turmoil that has engulfed the Federation recently. We have emerged, battered but still united. In the transition period when those who committed crimes will receive the system’s justice, the High Governor’s Board with my recommendation is appointing Alyie Starstriker to be the Interim High Governor. At her request, I will continue to serve as Vice High Governor. Now the High Justice will administer the Oath of Office.”

Alyie approached the dais. She was once again dressed in black, this time wearing a faux fur lined leather jacket for warmth against the cool mountain air over her turtleneck and a faux wool pencil skirt. To provide a balance with her hair color, she tucked her black trousers into tall burgundy boots with stacked heels.

High Justice Alana came to Alyie and recited. “Do you, Alyie Starstriker, swear to the best of your abilities to uphold the principles of the Sidereal Federation as High Governor and serve its members with fidelity, integrity, and justice?”

“I do.”

After waiting for the applause generated by her acceptance to cease, Alyie spoke,

“Citizens of the Sidereal Federation, the last time I addressed you, I pleaded for patience, and you responded. Now I ask for your trust as I am to assume a position I did not seek nor want. I promise to serve you until a new High Governor is elected. I will not run for office. My only reason for accepting the offer is my lifelong loyalty to the Federation. I will dedicate the time I have as Interim High Governor to reconcile differences and heal wounds throughout the Federation. My son, Tomus, was correct in saying that the images on the walls of Kanva have much to teach us. What images do we want other civilizations to know of us? I would hope we project a space-wide culture that celebrates diversity but one that cherishes the unity that preserves peace and stability. Thank you.”

After a second round of applause, Alyie stepped away and rejoined her group.

“Great speech, Mom. You really are getting good at this,” said Tessa, snidely smiling.

“So much for our retreat on Kanva,” said Jaynis. “Time to pack our bags again.”

“Hold on,” said Alyie, “with luck, this will be a short stay.”

“Be serious, Mother,” interjected Tomus. “Do you think that anyone can do this job better than you? My guess is that they will drag their feet setting up a new election with you in the High Governor’s seat.”

“Well, if so, that will affect you as well, Tomus.”

“What do you mean, Mother?”

“In my first act as Interim High Governor, I am appointing you my Chief of Staff. I saw how well you negotiated and helped delay a crisis. You will be my right-hand person.”

“What if I refuse?”

“You won’t, Tomus. Think of the lovely image on a wall of the High Governor and her Chief of Staff, mother and son, for a future civilization to see.”

Afterward

After Alyie's installation as Interim High Governor, all who were in attendance were shepherded to a reception in the Banquet Hall at Federation Earth Headquarters. At Alyie's request, it was a subdued affair due to the circumstances involving her ascension to the High Governor's seat. However, Jaynis, Tanita, and Alison Bernewithy had other ideas and made arrangements for a smaller but more exuberant celebration afterward.

As the reception in Colorado Springs was winding down, Alyie made the rounds accepting congratulations and saying her adieus to those who traveled from off planet. When that was all concluded, she discretely joined her family and their special friends and boarded a hypersonic airjet that Tessa would fly. On board were Alyie, Janis, Tanita, Asher, Tessa, Yong, Tomus, Melissa, and Alison. In less than two E-hrs. the plane landed in the New Brasilia spaceport where an airbus was waiting. On the landing field to greet them was Mr. Gustavo Sousa, an entrepreneur, and hotelier who was also a gentleman friend of Alison.

The 60-plus E-yr. old Gustavo was dressed elegantly in a shark's skin double-breasted suit with a violet ascot and wearing what looked like patent leather black shoes. His skin was bronze, and his wavy white colored hair was immaculately coiffured. As Alyie and her entourage approached him, Mr. Sousa stood at attention, greeting Alyie with "Olá" and then reaching out to kiss her on both cheeks.

A little taken aback by this familiarity, Alyie drew back, but Alison laughed, saying, "He does that to all the women he meets; it's a Brazilian thing."

After boarding the airbus, the party was whisked to Royal Tulip Hotel, which Gustavo owns, located right by the gorgeous Paranoá Lake. (A security patrol aboard a second airbus followed and coordinated with its counterpart at the hotel.) After refreshing themselves, the group was ushered into the dining room, where a festive meal was about to be served. On the flight, Alyie had changed out of her leather jacket and turtleneck into a silk burgundy halter top that wrapped around her neck, showing a good part of her clavicle and shoulders. Her skirt was circumscribed by a leather belt with a large silver buckle holding it snuggly in place.

The meal consisted of feijoada, white rice, cooked sweet cassava, and cheese bread. Everything was washed down by fruity sangria. As the wine took its effect, voices were raised, and laughter increased. In addition to Alyie and Tomus, there were two others seated at the gathering that were changing locations and positions. Tessa had been given her own fighter squadron to command. She named it Falcon and requested Emilie Aubere to be her co-commander. Melissa was given leave by Queen Revita to assume the position of Adjunct to Alison at Regional Headquarters in Brussels. The fact that this would provide proximity to Tomus in Colorado Springs did not escape notice from Alyie and Jaynis.

While the guests were drinking and eating, a samba band started playing. Alyie, never one to let the opportunity to dance go wasted, rose from her table with a glass of wine. Slightly inebriated, she called attention to the music, saying, "As High Governor, I declare this celebration would not be complete if we didn't all samba together. Shall we?"

As everyone, some more reluctantly than others, rose, Alyie called out, "Mr. Sousa, will you lead?"

Gustavo grabbed Alison, and they went to the space in front of the band and began the pulsating rhythmic movements. Alison was dressed in a short frilly yellow dress with a low plunging neckline and black high heels. Soon Alyie and Jaynis joined in, waving for the other couples to participate. Tanita and Asher were by far the most talented couple, showing style and grace while keeping in step with the band's rapid vibrations. Alyie was in high spirits, and she wheeled and pranced around Jaynis, gyrating her body toward and away from him. He latched on to her silver buckle as she wiggled her hips toward him. Tomus was not the most coordinated, but Melissa showed her skill and patience in guiding him in the basic steps. As for Tessa and Yong, the former did most of the twirling and prancing while the latter just stood in place.

After there was collective exhaustion, Gustavo gave the signal for the band to stop, and all retired out of breath to their seats. Having a final round of toasts before the tapioca cake and hot beverages were served, Mr. Sousa declared, "To our old and new High Governor. May her rule be forever."

To shouts of "Hurrah and Yes, Yes," Alyie raised her drink, which was somewhat wobbly in her hand, saying, "Mr. Sousa, is that a toast or a curse you just bestowed? Either way, let's all drink up."

After the frivolities were over, the guests retreated to their respective private rooms in the hotel: Alyie and Jaynis, Tanita and Asher, and Alison and Gustavo. Tessa and Yong and Tomus and Melissa were offered separate or coupled rooms, and each pair chose to bed down together.

Character List in Further Tales of Alyie Starstriker

Stories:

Reign of Alyie I (RAI)

Reign of Alyie (RAII)

Tessa and Tomus (TT)

The RIPEST Harvest (RH)

Images on the Wall I (IWI)

Images on the Wall II (IWII)

Afterward (A)

Alyie Starstriker and immediate family

Alyie Starstriker: High Governor of the Sidereal Federation (RAI, RAII, TT, RH, IWI, IWII, A).

Jaynis Bottombrook: Astrophysicist and eternal partner of Alyie Starstriker (RAI, RAII, TT, RH, IWI, IWII, A).

Tessa Starbrook: Daughter of Alyie and Jaynis and twin of Tomus (RAI, RAII, TT, RH, IWI, IWII, A).

Tomus Starbrook: Son of Alyie and Jaynis and twin of Tessa (RAI, RAII, TT, RH, IWI, IWII, A).

Tanita Starbright: Alyie Starstriker's mother and former Sidereal Federation Space Commander (RAI, RAII, TT, RH, IWI, IWII, A).

Alex Striker: Alyie's Starstriker's father and tactician for Space Command (RAI, RAII).

Talia Silverfoot: Mother of Jaynis Bottombrook (RH).

Bernice Bottombrook: Sister to Jaynis Bottombrook (RAII, RH).

Saada Nakawa: Life partner of Bernice Bottombrook (RAII).

Valkyrie Squadron Members: Olga Fercouska (RAII), Hollie Pederson (RAII), Zarah Khoury (RAII) and Wang Li (RAII).

Sidereal Federation Officials

Steely Cranshaw: Admiral in Space Command and High Council Member (RAI, RAII, RH).

Komesh Thakur: High Council Member and High Governor (DC, IWII).

Victoria Alana: High Justice (IWII).

Bristol Menhavi: Vice High Governor (IWI, IWII).

Rufus Seul: High Governor's Board Member (IWII).

Ivan Presoff: General and Head of Ground Forces (RAI, RAII).

Sonja Benita: Director of Federation Scientific Institute (RAI).

Ja-Li Wang: Engineer at Federation Safety and Security Lab. (RAI).

Lenora Honeywine: Captain and Co-Commander of *Argonaut* (RAI, RAII).

Ronica Richlieu: Deputy High Governor (RAI).

Daiyu Fong: Director of Common Exchange (RAI, RAII).

Alphonse Cache: Cofounder of Cache Enterprises (RAI, RAII).

Fernando Alvarez: Deputy High Governor (RAII).

Curt Beckwall: Brigadier Commander (RAII).

Arbrit Margiven: Colonel and Superintendent of Fleet Academy (TT).

Finneas O'Rourke: Officer at Department of Security (TT, IWII).

Theodore Ramsey: Captain of Panthera (TT).

Viola Narumbi: Lieutenant of Panthera (TT).

Naja Kwan: Fleet Academy Cadet (TT).

Argus Fitzhugh: Fleet Academy Cadet (TT).

Objie Narubi: Fleet Academy Cadet (TT).

Rufus Steinhaus: Terra Engineer and *Seleucus* member (RH).

Cybele Wainforth: Planetary Scientist (RH).

Singha Khan: Captain and Commander of *Hesperus* (IWI).

Grazia Alvarez: Lieutenant and Co-commander of *Hesperus* (IWI).

Emilie Aubere: Lieutenant on *Hesperus* (IWI, IWII).

Bryan Doherty: Ensign on *Hesperus* (IWI, IWII).

Stanis Aboru: Ensign on *Hesperus* (IWII).

Albus Decker: First Mate on *Hesperus* (IWII).

Yamuchi Osaka: Admiral and Commander of *Olympiad* (IWII).

Camilla Fuselli: Lieutenant at Safety and Security Office (IWII).

Anya Geranova: Officer, Second Rank at Safety and Security Office (IWII).

Joseph Kreiger: Major under General Reg Jones (IWII).

Omar Fanzi: Lieutenant in Fleet Command (IWII).

Taneous Taub: Captain and Commander of *Dreadnought*: (IWII).

Liu Fan: Exobiologist (RH).

Mahmoud Foustak: Space Systems Engineer (RH).

On Planet Ergo

Gianni Vincenti: Trade Guild Leader (IWII).

Rosetta Vincenti: wife of Gianni (IWII).

Diego Vincenti: son of Gianni and Rosetta (IWII).

Conchetta Vincenti: daughter of Gianni (IWII).

Stephen Marco: Director of Information Center (IWII).

On Earth

Hector Palermo: Regional Governor (RAI).

Silvia Gonzalez: Premier on Earth (RAII).

Hashin Samak: Deputy Director of Ministry of Arts and Culture (IWI).

Renard Fouya: Professor of Hieroglyphologoly at Université de Paris (IWI).

Gustavo Sousa: Entrepreneur and Hotelier (A).

On Kanva

Revita Jadeite: Queen of Kanva (RAI, RAII, IWI).

Rebina Ectot: Lady-in-waiting to Queen Revita (RAI).

Melissa Turnberry: Lady-in-waiting to Queen Revita (IWI, IWII, A).

Samuel Kohn: Exobiologist (RAI).

Mai Pham: Exobiologist (RAI).

Devon Ricketts: Linguist (RAI).

Epson Alandro: Parliamentary Leader (RAI).

Ruha Sonoa: Medical Doctor (RAI).

Greta Gudwig: Attendant to Alyie Starstriker (IWII).

Other Characters

Reg Jones: General and Vice High Governor (RAI, RAII, TT, RH, IWI, IWII).

Alison Bernewithy: Director of Arts and Culture Ministry and Regional Governor (RAI, RAII, TT, RH, IWI, IWII, A).

Brianna Crosby: Lieutenant and Adjunct to Reg Jones (RAII, TT).

Fedor Greysan: Pilot of Bellerophon (RAII).

Alphonse Cache: Head of Cache Enterprises (RAI, RAII).

Yong Gwan: Planetary Scientist and Chief Mission Scientist for RIPEST (RH, IWII, A).

Anwar Mulafta: Planetary Specialist on *Seleucus* (RH).

Millicent Dubwa: Atmospheric Scientist on *Seleucus* (RH).

Nyra Fatima: Hydrologist on *Seleucus* (RH).

Lucca Comani: Captain and Commander of *Seleucus* (RH).

Marisa Gomez: Physician and Medic on *Seleucus* (RH).

B?alamians

Thu?glr Kap?stk: Captain and Commander of Alux (RAI, RAII).

Glib D?rhim: First Navigation Officer of Alux (RAI).

Ben?ib Akal?een: First Communication Officer of Alux (RAI, RAII).

Hi?nigh Selm?dom: Alien Technology Officer of Alux (RAI).

Roi Sing?a XV: High Exalted One of B?alam (RAI, RAII).

Roi Leonidas VII: Exalted Ruler of B?alam (RAII).

On Cognito

Asher Weinbein: Hotelier of the Hallow Inn (TT, RH, IWI, IWII, A).

Carlyle Denton: Staff member at the Hallows Inn (TT).

Nina Mori: Staff member at the Hallows Inn (TT).

Made in the USA
Monee, IL
04 November 2026

67378727R00197